The Rural Landscape of Ancient Israel

Edited by

Aren M. Maeir
Shimon Dar
Ze'ev Safrai

BAR International Series 1121
2003

Published in 2016 by
BAR Publishing, Oxford

BAR International Series 1121

The Rural Landscape of Ancient Israel

ISBN 978 1 84171 497 4

BAR Publishing is the trading name of British Archaeological Reports (Oxford) Ltd. British Archaeological Reports was first incorporated in 1974 to publish the BAR Series, International and British. In 1992 Hadrian Books Ltd became part of the BAR group. This volume was originally published by Archaeopress in conjunction with British Archaeological Reports (Oxford) Ltd / Hadrian Books Ltd, the Series principal publisher, in 2003. This present volume is published by BAR Publishing, 2016.

Printed in England

BAR
PUBLISHING

BAR titles are available from:

BAR Publishing
122 Banbury Rd, Oxford, OX2 7BP, UK
EMAIL info@barpublishing.com
PHONE +44 (0)1865 310431
FAX +44 (0)1865 316916
www.barpublishing.com

Table of Contents

List of Illustrations

Foreword

This volume owes its origins to a collection of studies entitled *The Village in Ancient Israel* (Dar and Safrai 1997), which included papers (in Hebrew) that dealt with a selection of topics relating to village life in the Land of Israel during various periods. Following the publication of that volume, the editors (Dar and Safrai) felt the need for a similar volume in the English language. And it was thus that the third editor (Maeir) of the present volume joined with them to put out an English language collection.

It soon became apparent that the scope of the original volume had to be changed. Thus, a broader range of study was chosen for the new volume, in which an attempt would be made to collect studies that dealt with aspects of rural life from as wide a range as possible, and from as many periods as possible. In addition, it was decided that all the articles would be previously unpublished contributions, and not English translations of the Hebrew studies published in the earlier volume. It was hoped that this would offer an opportunity to view a wide range of perspectives on the non-urban culture in the Land of Israel throughout many periods.

The potential authors were invited to contribute papers to this volume based on the following criteria:

1) The studies in the collection would deal with various synthetic aspects of the rural way-of-life in the ancient Land of Israel, from various historical and archaeological perspectives.
2) The temporal range of the studies would span the late Prehistoric through Modern periods.
3) The various contributions would deal with general phenomena, though they may be limited to specific periods or geographic regions. Thus, the discussions would not deal with a single site or installation, but rather discuss a type of settlement or agricultural installation, the settlement in a given region or during a given period/s, etc.
4) The volume would focus on various topics, including *Physical Makeup* (such as: architecture, village planning, inter-village installations, etc.), *Rural Settlement Patterns and Processes*, and *Rural Economic Activities* (such as: agriculture, production and trade).

All told, ten authors contributed studies to this volume, studies that cover a wide range of topics and perspective, and in fact, do touch on many of the original goals set for the volume. We would like to thank all the authors for their contributions, and for their patience for the over-extended timetable of publication.

The editors would like to thank Ms. Edna Sachar for her meticulous copy-editing, and Mr. Joe Uziel for serving as the volume coordinator and for assiduously arranging the camera-ready outlay of the volume. Likewise, thanks to the Kushitzky Foundation for supporting the publication of the volume.

The Editors,
Ramat Gan
January, 2003

Introduction: The Rural Landscape of Ancient Israel

AREN M. MAEIR, SHIMON DAR, and ZE'EV SAFRAI

The Institute of Archaeology
The Martin (Szusz) Department
of the Land of Israel Studies
and Archaeology
Bar Ilan University
Ramat Gan, 52900
Israel

The rural components of various ancient societies have been the focus of much research in recent years. To a large extent, this trend is the result of a reaction against a purported emphasis on the study of the urban facets of society (and related perspectives) in most traditional, "normative" historical and archaeological research. Accordingly, due to the priority given to the excavation of stratified tells and other large sites, as well as the interest in monumental archaeological remains and geo-political historical perspectives, the study of "less complex" but ubiquitous village life has fallen by the wayside in most research agendas (see further, Ahlström 1982; London 1989; Liverani 1999).

To counter this situation, numerous scholars have focused their attention on the rural, non-urban aspects of the study of ancient societies. This is seen both in the research conducted on the early, formative, and pre-urban stages of human history (e.g., Flannery 1976; Tchernov 1994), as well as in studies covering virtually all the later stages of human culture. In effect, although one often has the impression that mainstream archaeology is still "urban-oriented," a tremendous volume of research has appeared on rural-related issues. For example, archaeological studies and excavation reports on rural sites (e.g., Dar, Tepper, and Safrai 1986; Seeden 1986; Ben-Tor et al. 1987; Falconer and Magness-Gardiner 1989; Lehmann et al. 1991; Dar 1999; Dessel 1999) and social-historical and/or ethnoarchaeological studies of villages (e.g., Bates 1980; Gilbert 1982; Biger and Grossman 1992; Grossman 1994; Horne 1994). Nevertheless, a certain lack of broad-ranging overviews and wider perspectives was felt, particularly in relation to the archaeology of the ancient Near East, and several studies, as well as collections of papers have attempted to bridge this gap (e.g., Schwartz and Falconer 1994; Faust 2000).

In this context, two of the editors of this volume, both of whom had previously studied aspects of the rural history and archaeology of the Land of Israel (e.g., Dar et al. 1986; Safrai 1998; Dar 1999), decided to bring together a collection of rural-related studies, and published an edited volume in which various rural perspectives relating to different periods in ancient Palestine were presented in Hebrew (Dar and Safrai 1997). Originally, the present volume was planned as a translation of the former. Eventually, however, the three editors decided that a completely new collection of rural-oriented studies was preferable, in order to be able both to add more such articles and to include non-Hebrew contributions in the volume.

The title of this volume, *The Rural Landscape of Ancient Israel*, needs some clarification, due to the often nebulous and at times multivalent terms used therein. The choice of the term "Rural" is quite straightforward, indicating all aspects of non-urban, and mostly sedentary, lifestyles. This includes such facets as settlement pattern and distribution, site classification, social structure, economic and agricultural foundations, cultic manifestations, legal definitions, and human-animal relations.

The term "Landscape" is somewhat more problematic. This term is used (and in some cases, misused) in a wide variety of contexts (for overviews see, e.g., Knapp 1997:14-18; Anschuetz, Wilshusen, and Scheik 2001). Originating primarily in the realm of art and art history (e.g., Schama 1995; Warnke 1995), it is now commonly used in geographical (e.g., Cosgrove 1985), historical (e.g., Perry 1999), anthropological (e.g. Hirsch and O'Hanlon 1995), and archaeological studies (see below).

There has been a marked proliferation in the archaeological literature in the use of the term "landscape," so much so that in recent years, numerous attempts have been made to present methodological and theoretical overviews (e.g., Ingold 1993; Knapp 1997; Johnston 1998; Huot 1999; Knapp and Ashmore 1999; Anschuetz, Wilshusen, and Scheik 2001; Smith 2001; Hicks 2002; Finlayson and Dennis 2002) and numerous collections of studies have appeared (e.g., Crumley 1994; Ashmore and Knapp 1999; Milano et al. 1999a; 1999b; 2000; Ucko and Layton 1999). Many of these (and other) discussions have noted the varied meanings (often, but not always, complementary) of the term landscape. In order to simplify the definition, a distinction is usually made between the use of landscape as signifying the "natural environment" (quite similar to "environment"), as opposed to landscape denoting the cultural and ideological concepts of the natural and cultural surroundings of any given society at any given time (e.g., Ingold 1993; Huot 1999; Anschuetz,

Wilshusen, and Scheik 2001). Thus, the term has an enormously wide definition and subsequent use.

In this volume, the broad view of the term landscape is applied. Thus, in referring to the "Rural Landscape," any aspect of the rural surroundings of human culture are included, whether phenomena of natural or manmade (physical and/or ideological) origin.

As for the term "Ancient Israel," although this is without doubt an ideologically highly loaded term, in this case it is used simply to refer to the pre-modern periods within the approximate region of the modern-day State of Israel.

Although it was hoped originally that a larger number of studies would be included in this volume, the 10 contributions deal with an impressively wide range of topics, periods, and methodological approaches. While, needless to say, these studies cover but a small portion of the possible topics included under the heading "The Rural Landscape of Ancient Israel," they offer a good cross-section of the varied approaches to selected aspects within this broader horizon.

Following this Introduction, S. Gibson presents an overview of the theoretical, methodological, and practical underpinnings of "Landscape Archaeology," basing his discussion on the results from several such projects that he himself has directed. Given that much of the basic information for the study of the rural landscape is to be derived from just such studies, Gibson's comments, based on years of field experience, are well-placed. The successful implementation of the methods he describes, as well as their apt correlation with other archaeological research results, is a fine example of the utility of such approaches.

There is though one aspect of Gibson's paper that we would like to qualify. From reading his contribution, one can get the impression that prior to recent times, and in fact, save for his own studies, little, if any "Landscape Archaeology" was conducted in the Land of Israel. This is far from the truth. More so, from very early stages of the archaeological research of the land, and continuing to the present-day research, many scholars have chosen field methodologies and interpretive approaches that, even if not dubbed as such, are in fact "Landscape Archaeology", or at the very least, methodologically comparable. Several examples will suffice. For earlier researchers, to a large extent, the studies of Reifenberg (1955), Kedar (1967), and Evenari et al. (1982) on the relationship between man and nature in the arid environments serve as fine examples of early efforts. From more recent times, studies by M. Haiman (e.g., 1992a; 1992b), R. Frankel (e.g., Frankel et al. 2002), and one of the editors of this volume, Dar (e.g., 1986), to name but a few, demonstrate that an awareness to such approaches, and the appropriate "mindset" for the implementation of such methodologies did exist, even if it was not explicitly labeled "Landscape Archaeology"

R. Greenberg's contribution focuses on rural settlement during the Early Bronze Age, with particular emphasis on the Hula Valley. While concentrating on rural settlement, Greenberg takes to task some recent views that appear to over-accentuate the importance and predominance of the rural component in the settlement pattern in the ancient Levant. He critiques the approach of several scholars, such as S. Falconer (1987), A. Joffe (1993), and G. Palumbo (1990), who have posited that the rural component should be seen as the constant in ancient Levantine society, with occasional "outbursts" of urban phenomena. Greenberg demonstrates both that the rural component was not a "constant" in the Levantine landscape, but underwent substantial vicissitudes, and that, similarly, the appearance of urban entities cannot be seen as evolutionary outgrowths of earlier rural settlements, but in fact developed for a wide range of reasons (often of non-local origin).

After a hiatus of several years in which he has not published studies relating to the Early Bronze Age IV (EB IV), W.G. Dever, building and expanding on both his own and others' previous work, attempts to map out the ideological underpinnings behind the settlement pattern of this period. Going beyond the standard definition of the period, as one in which pastoral and rural lifestyles are dominant, Dever, in the only study in this volume that deals in a full-fledged manner with the ideational aspect of the archaeological landscape, attempts to understand the settlement pattern in the EB IV as echoing the "mental map" and ideological preferences of the population during this period. He thus posits that the marked changes seen during the EB IV (and the virtual disappearance of urban society) are not solely the result of the contemporary existential necessities, but that the ideological reasoning and justification behind these changes in EB IV society must be considered as well.

A.M. Maeir, in discussing the relationship between the rural and urban components of the settlement landscape during the Middle Bronze Age II (MB II), addresses, like R. Greenberg, viewpoints that seem to have over-emphasized the importance of the rural settlements within the settlement matrix of this period. He critiques the suggestion put forth by S. Falconer (1987) that there were few, if any, urban settlements in the southern Levant during virtually all the pre-Hellenistic periods. Focusing on the MB II evidence from the Jordan Valley, Maeir demonstrates that urban entities did in fact exist. This is clear both from an analysis of the relative size (and functional significance) of the various MB II sites and from the proposed definition of this period's "sacred landscape" (which, to a certain extent, attempts to read the ideological underpinning behind the locations of choice cultic sites). The end result is a clear demonstration that the settlement matrix during this period, as well as during the other Bronze Age and the Iron Age periods in the Levant, is comprised of both urban and rural components, which are complementary and interdependent. Attempts to overstate the role of one at the expense of the other leads to a mistaken understanding of the social, economic, and geographic realities in Ancient Israel.

Over the past few decades, numerous attempts have been made to explain the underlying mechanisms behind the sudden appearance and subsequent abrupt disappearance of the several hundred rural sites in the Northern Negev, in the course of the early Iron Age II (ca. 10th century BCE). M. Haiman, after summarizing the relevant data, demonstrates the similarities between the sites in the Negev and rural sites in the more temperate zones of Palestine, indicating a common origin. He argues that, consequently, the only acceptable rationale behind the appearance of these settlements in the Negev is to view them as an extension of the economic policy of a polity centered in the temperate zone, which he believes can be identified with the "United Monarchy" in Judah. Conversely, their disappearance should be associated with the military campaign of Sheshonq/Shishak, one of the objectives of which was to restrict the economic and military dimensions of the "United Monarchy." All told, Haiman ably demonstrates the intimate connections that existed between the rural and urban components and the temperate and arid zones in the Iron Age settlement landscape. It should be borne in mind, however, that in light of recent suggestions to lower the chronology of the Iron Age (e.g. Finkelstein 1996) and alternative historical reconstructions (e.g. Na'aman 1998), other explanations for this phenomenon are also possible.

A. Faust, in a study of rural settlements in Ancient Israel during the Iron Age, defines the farmstead as a distinctive type of rural settlement. Differentiating it from the village, which likewise existed during this period, and equating it with the biblical term ha□er, he believes that the farmstead represents the abode and economic mainstay of extended families. In addition, he points out that the farmstead appears to be more common in the southern Kingdom of Judah than in the northern Kingdom of Israel, which he believes reflects different levels of social and economic development. This definition is of considerable importance, since (following Stager's seminal work [1985]), it moves beyond a functional and/or size-oriented typology of Iron Age rural settlements, towards an understanding of the social and kinship-based underlying frameworks.

In a paper covering several periods, Z. Safrai attempts to classify and legally define the various types of extra-mural lands in ancient Palestine. Focusing primarily on the Roman and Byzantine periods (ca. 60 BCE-630 CE), but reflecting on earlier and later periods as well, Safrai analyzes both Jewish and Roman legal sources, as well as epigraphic and other data, delineating several different classes of land and reconstructing the agrarian infrastructure during these periods.

J. Schwartz discusses a little-known topic, namely, the role of the domesticated dog in Jewish rural life during the Roman and Byzantine periods. Based on the historical and legal sources, he demonstrates the various and, one might add, somewhat surprisingly important roles that these dogs fulfilled.

The two centuries of Frankish rule in the Levant (the Crusader period, 1099-1291 CE) heralded the appearance of numerous new and foreign cultural elements in Palestine. As expected, this is manifested in the rural realm as well, as A.J. Boas demonstrates in his examination of several distinctive types of rural settlement unique to this period. Discussing the appearance of the "rural estate," the farmhouse, and the "linear village," Boas traces their origins and social and economic roles within Frankish society.

In the final paper, Z. Amar deals with the latest period treated in this volume, namely, the Mamluk period. Discussing the agricultural infrastructure and governmental policy during the Mamluk period, Amar attempts to document, define, and understand the process of agricultural decline so clearly apparent during this period. He demonstrates that although this decline was obviously the result of extremely faulty economic and social policies of the Mamluk rulers, it should also be seen within the context of contemporaneous agricultural and technological developments in Europe, which only accelerated the local process of decline.

In closing, it is hoped that the studies presented in this volume not only serve as interesting case studies of the rural components brought to the forefront of the historical and archaeological research of Ancient Israel, but demonstrate, once again, the general utility and potential of such research.

REFERENCES

Ahlström, G.W.
1982 Where Did the Israelites Live? *Journal of Near Eastern Studies* 42: 133-38.

Anschuetz, K., Wilshusen, R., and Scheik, C.
2001 An Archaeology of Landscapes: Perspectives and Directions. *Journal of Archaeological Research* 9/2: 157-211.

Ashmore, W., and Knapp, A.B., eds.
1999 *Archaeologies of Landscape: Contemporary Perspectives. Social Archaeology*. London: Blackwell.

Bates, D.
1980 The Middle Eastern Village in Regional Perspective. Pp. 161-83 in *Village Variability in Contemporary Society*, eds. P. Reining and B. Lenkerd. American Association for the

Advancement of Science Selected Symposium 34. Boulder: Westview.

Ben-Tor, A., et al.
1987 *Tell Qiri, a Village in the Jezreel Valley: Report of the Archaeological Excavations 1975-1977.* Qedem 24. Jerusalem: Hebrew University.

Biger, G., and Grossman, D.
1992 Population Density in the Traditional Village of Palestine. *Cathedra* 63: 108-21, 188 (Hebrew, with English summary).

Cosgrove, D.
1985 Prospect, Perspective and the Evolution of the Landscape Idea. *Transactions of the Institute of British Geographers* 10: 45-62.

Crumley, C., ed.
1994 *Historical Ecology: Cultural Knowledge and Changing Landscapes.* School of American Research Advanced Seminar Series 5. Santa Fe: School of American Research.

Dar, S.
1986 *Landscape and Pattern: An Archaeological Survey of Samaria, 800 B.C.E.-636 C.E.* British Archaeological Reports International Series 308. Oxford: British Archaeological Reports.
1999 *Sumaqa: A Roman and Byzantine Jewish Village on Mount Carmel, Israel.* British Archaeological Reports International Series 815. Oxford: BAR Publishing.

Dar, S., and Safrai, Z., eds.
1997 *The Village in Ancient Israel.* Tel Aviv: Eretz (Hebrew).

Dar, S., Tepper, Y. and Safrai, Z.
1986 *Um Rihan: A Village of the Mishnah.* Tel Aviv: Hakibbutz Hameuchad (Hebrew).

Dessel, J.P.
1999 Tell 'Ein Zippori and the Lower Galilee in the Late Bronze and Iron Ages: A Village Perspective. Pp. 1-32 in *Galilee through the Centuries: Confluence of Cultures*, ed. E.M. Meyers. Winona Lake, IN: Eisenbrauns.

Falconer, S.E.
1987 *Heartland of Villages: Reconsidering Early Urbanism in the Southern Levant.* PhD Dissertation. Tucson: University of Arizona.

Falconer, S.E., and Magness-Gardiner, B.
1989 Bronze Age Village Life in the Jordan Valley: Archaeological Investigations at Tell el-Hayyat and Tell Abu en-Ni'aj. *National Geographic Research* 5/3: 335-47.

Faust, A.
2000 The Rural Community in Ancient Israel during Iron Age II. *Bulletin of the American Schools of Oriental Research* 317: 17-39.

Finkelstein, I.
1996 The Archaeology of the United Monarchy: An Alternative View. *Levant* 28: 177-87.

Finlayson, B., and Dennis, S.
2002 Landscape, Archaeology and Heritage. *Levant* 34: 219-27.

Flannery, K.V., ed..
1994 *The Early Mesoamerican Village. Studies in Archaeology.* Orlando: Academic.

Gilbert, J.
1982 Rural Theory, the Grounding of Rural Sociology. *Rural Sociology* 47: 609-33.

Grossman, D.
1994 *Expansion and Desertion: The Arab Village and Its Offshoots in Ottoman Palestine.* Jerusalem: Yad Izhaq Ben-Zvi (Hebrew).

Hicks, R.
2002 Ways of Inhabiting the World: Landscape Archaeology. *American Anthropologist* 104/1: 315-19.

Hirsch, E., and O'Hanlon, M., eds.
1995 *The Anthropology of Landscape: Perspectives on Place and Space. Oxford Studies in Social and Cultural Anthropology, Cultural Forms.* Oxford: Clarendon.

Horne, L.
1994 *Village Spaces: Settlement and Society in Northeastern Iran.* Smithsonian Series in Archaeological Inquiry. Washington, DC: Smithsonian Institution.

Huot, J.-L.
1999 The Archaeology of Landscape. Pp. 29-36 in *Landscapes: Territories, Frontiers and Horizons in the Ancient Near East.* Papers Presented to the XVIL Rencontre Assyriologique Internationale, Venezia, 7-11 July 1997. Part I: Invited Lectures, eds. L. Milano et al. History of the Ancient Near East Monograph III/1. Padova: Sargon.

Ingold, T.
1993 The Temporality of Landscape. *World Archaeology* 25: 152-74.

Joffe, A..
1994 *Settlement and Society in the Early Bronze I & II Southern Levant: Complementarity and Contradiction in a Small-Scale Complex Society.* Monographs in Mediterranean Archaeology 4. Sheffield: Sheffield Academic.

Johnston, R.
1998 Approaches to the Perception of Landscape: Philosophy, Theory, Methodology. *Archaeological Dialogues* 5/1: 54-68.

Knapp, A.B.
1997 *The Archaeology of Late Bronze Age Cypriot Society: The Study of Settlement, Survey, and Landscape.* Department of Archaeology Occasional Paper 4. Glasgow: University of Glasgow.

Knapp, A.B., and Ashmore, W.
1999 Ideational and Industrial Landscape on Prehistoric Cyprus. Archaeological Landscapes:

Constructed, Conceptualized, Ideational. Pp. 1-30 in *Archaeologies of Landscape: Contemporary Perspectives*, eds. W. Ashmore and A.B. Knapp. Social Archaeology. London: Blackwell.

Lehmann, G., et al.
1991 The 1990 Excavations at Abu Snesleh: Preliminary Report. *Annual of the Department of Antiquities of Jordan* 35: 41-62.

Liverani, M.
1999 The Role of the Village in Shaping the Ancient Near Eastern Rural Landscape. Pp. 37-48 in *Landscapes: Territories, Frontiers and Horizons in the Ancient Near East*. Papers Presented to the XVIL Rencontre Assyriologique Internationale, Venezia, 7-11 July 1997. Part I: Invited Lectures, eds. L. Milano et al. History of the Ancient Near East Monograph III/1. Padova: Sargon.

London, G.
1989 A Comparison of Two Contemporaneous Lifestyles of the Late Second Millennium B.C. *Bulletin of the American Schools of Oriental Research* 273: 37-55.

Milano, L., et al., eds.
1999a *Landscapes: Territories, Frontiers and Horizons in the Ancient Near East*. Papers Presented to the XVIL Rencontre Assyriologique Internationale, Venezia, 7-11 July 1997. Part I: Invited Lectures. History of the Ancient Near East Monograph III/1. Padova: Sargon.
1999b *Landscapes: Territories, Frontiers and Horizons in the Ancient Near East*. Papers Presented to the XVIL Rencontre Assyriologique Internationale, Venezia, 7-11 July 1997. Part II: Geography and Cultural Landscapes. History of the Ancient Near East Monograph III/2. Padova: Sargon.
2000 *Landscapes: Territories, Frontiers and Horizons in the Ancient Near East*. Papers Presented to the XVIL Rencontre Assyriologique Internationale, Venezia, 7-11 July 1997. Part III: Landscape in Ideology, Religion, Literature and Arts. History of the Ancient Near East Monograph III/3. Padova: Sargon.

Na'aman, N.
1998 Shishak's Campaign to Palestine as Reflected by the Epigraphic, Biblical and Archaeological Sources. *Zion* 63/3: 247-76 (Hebrew, with English abstract).

Palumbo, G.
1990 *The Early Bronze Age IV in the Southern Levant: Settlement Patterns, Economy, and Material Culture of a 'Dark Age'*. Contributi e Materiali di Archaeologia Orientale III. Rome: Universitá Degli Studi di Roma La Sapienza.

Perry, W.
1999 *Landscape Transformation and the Archaeology of Impact: Social Disruption and State Formation in Southern Africa*. New York: Plenum.

Safrai, Z.
1998 Ancient Field Structures: The Village in Eretz Israel during the Roman Period. *Cathedra* 89: 7-40 (Hebrew, with English abstract).

Schama, S.
1995 *Landscape and Memory*. New York: Vantage.

Schwartz, G.M., and Falconer, S.E., eds.
1994 *Archaeological Views from the Countryside: Village Communities in Early Complex Societies*. Washington, D.C.: Smithsonian Institution.

Seeden, H.
1986 Bronze Age Village Occupation at Buṣrā: AUB Excavations on the Northwest Tell 1983-84. *Berytus* 34: 11-81.

Smith, A.
2001 On Landscapes in the Ancient Near East. *Journal for the Study of the Economic and Social History of the Orient* 44/3: 363-71.

Stager, L.E.
1985 The Archaeology of the Family in Ancient Israel. *Bulletin of the American Schools of Oriental Research* 260: 1-35.

Tchernov, E.
1994 *An Early Neolithic Village in the Jordan Valley, Part II: the Fauna of Netiv Hagdud*. American School of Prehistoric Research, Bulletin 44. Cambridge, MA: Peabody Museum.

Ucko, P., and Layton, R., eds.
1999 *The Archaeology and Anthropology of Landscape: Shaping Your Landscape*. London: Routledge.

Warnke, M.
1995 *Political Landscape: The Art History of Nature*. Cambridge, MA: Harvard University.

From Wildscape to Landscape: Landscape Archaeology in the Southern Levant – Methods and Practice

SHIMON GIBSON

W.F. Albright Institute of Archaeological Research
P.O. Box 19096
Jerusalem, 91190
Israel
shimgib@yahoo.com

It is suggested that the archaeological study of the chronological and spatial transformation of open areas in the southern Levant from "wildscapes" into rural and industrial landscapes, could be significantly facilitated and advanced by employing basic procedures of Landscape Archaeology similar to those currently being used in other parts of the world. Various methods of landscape investigation are discussed, notably field surveys, regional surveys and site-catchment analysis. The strengths inherent in the methods and practice of Landscape Archaeology are emphasized, particularly for researchers dealing with the investigation of the morphological variables of ancient settlement patterns and the physical remains of agricultural regimes. Examples are given of projects that have successfully employed Landscape Archaeology methods as part of their research design.

Introduction

The investigation of ancient landscapes would seem at face value to be a subject that should be a natural and integral part of all archaeological research programmes in the southern Levant (Israel and Palestine). This is not in fact the case and, until fairly recently, the subject has been ignored, with archaeological projects concentrating almost exclusively on the investigation of various forms of ancient settlements -- from tells to small villages -- and research dealing with aspects of architecture and material culture. In contrast, the archaeological investigation of the development of landscapes has flourished in England and Europe since the 1970s and Landscape Archaeology has attained a very honourable position in the curricula of many of their universities.

While Landscape Archaeology has had a significant impact in Mediterranean countries (notably in Spain, Italy and Greece), Biblical Archaeology has dominated the archaeology of the southern Levant and this has had a fundamental effect on the direction modern archaeology has taken. As a result, Israel has become one of the most excavated of countries in the world and an enormous amount of light has been shed on the archaeology of the biblical periods. In the past decade, however, some of the limitations of Biblical Archaeology have become glaringly evident and younger generations of archaeologists are turning away from this traditional approach and becoming much more interested in investigating alternative methods of archaeological research. Consequently, fieldwork methods are becoming more scientific and interest in the investigation of the archaeological remains of non-biblical periods has increased. There is also a greater interest in theoretical approaches (such as the cognitive and post-processual approaches) and in the benefits of multi-disciplinary work, and field surveys have become more professional. However, the proper analytical investigation of ancient landscapes ("Landscape Archaeology") has yet to make an impact in Israel and Palestine, firstly, as a major component of the research design of archaeological projects and, secondly, as an integral part of the teaching programmes of Israeli and Palestinian universities.[1]

A Definition of Landscape Archaeology

Landscape Archaeology is an all-inclusive and highly flexible method for studying the development through time and space of a continuous distribution of large and small man-made features across a given landscape. In using this method, one is attempting to explain how and what one sees today came to look the way that it does and to interpret the spatial patterns and structures created in the past in terms of social and economic behaviour. The method is particularly suitable for investigating rural landscapes -- for example, farms with their field systems or areas of terraced hill-slopes -- which have tended to defy analysis and interpretation using conventional archaeological methodologies (Fig. 1-1). In this method, the entire landscape is in effect the "site" and settlements are only features within the overall landscape, albeit important features, since they represent foci of human activities. This definition of Landscape Archaeology as a form of "total archaeology" -- settlement patterns, field systems, territories and communications -- differs from the conventional concept of "off-site" features spread *between* occupation sites. Defining the entire landscape (or the part chosen for research) as the "site" is far more expansive than the restricted geographical definition of the term as reflecting the location or situation of a place (Cain 1963: 307-8; Wagstaff 1991: 9).[2]

The origins of Landscape Archaeology in England, may be traced back to the pioneering survey methodology, using aerial photography, developed by O. G. S. Crawford in the 1930s. His classic book *Archaeology in the Field* (1953), was a landmark in the study of landscape history. Crawford regarded the landscape as a palimpsest, "...like a document that has been written on and erased over and over again; and it is the business of the field archaeologist to decipher it." (1953: 51) Another study, *Ancient Landscapes: Studies in Field Archaeology* (1957), by John Bradford was also extremely influential in the subsequent development of Landscape Archaeology. By now, Landscape Archaeology has become an integral part of archaeological research programs at universities and widely used by field units working in England and different parts of Europe (Aston and Rowley 1974; Aston 1985; Roberts 1987). Methods of Landscape Archaeology have also been adopted as an integral part of archaeological projects carried out in various Mediterranean countries (Keller and Rupp 1983; Macready and Thompson 1985; Van Andel, Runnels and Pope 1986; Van Andel and Runnels 1987; Barker and Lloyd 1991; Barker and Jones 1985; Cherry, Davis and Mantzourani 1991) (Fig. 1-2).

Landscape interpretation overlaps with and draws upon the work of historians, historical geographers, ethnographers, historical ecologists and others. New theoretical approaches to the subject are examining manifestations of recursive relationships between the cultural landscape (at various scales), social action and perceptions of the world (Rossignol and Wandsnider 1992; Miller and Gleason 1994; Ucko and Layton 1999).

Methods of Landscape Archaeology

While the philosophy and basic principals of Landscape Archaeology remain generally the same from country to country, methods of application and emphasis can differ considerably from one project to another even within the same country. This is evident from the publications on the subject. Hence, methods used in England, for instance, should not be imported wholesale to Israel/Palestine.[3] Instead, there must be flexibility in the design of Landscape Archaeology projects based on the research goals of individual archaeologists.

The general failing of the methods previously used for the examination of landscapes, namely field surveys, regional surveys and site catchment analysis, has lain in the overall rigidity of their application. The conventional field survey is concerned primarily with locating "sites", i.e. places showing evidence of human occupation usually represented by architectural remains and pottery.[4] This is the method used by the Archaeological Survey of Israel (now a division of the Israel Antiquities Authority).[5] A site could be represented either by a sherd scatter or a tel. The major drawback of this method is that sites are not investigated within their environmental boundaries but rather within the artificial boundaries of 10 x 10 kilometre maps. The landscape is almost totally ignored and no attempt is made to examine field systems unless they are clearly defined and belong to single-period settlements (this is usually possible only in the Negev Highlands and Judean Desert). No attempt is made to look at the *pattern* of "off-site" features within a given landscape (such as wine and oil presses, cisterns, threshing floors, terraces, fields, quarries or roads) and these are merely listed. In the end, all that can be derived from this method are maps of settlements and find spots, and only very little information regarding the development of the landscape as a whole.[6]

The regional survey is a more advanced method than the conventional field survey, but the "site" or settlement remains the primary unit of analysis and the tendency is to ignore "off-site" features as a whole, regarding them as "background noise" of no significance for general settlement patterns (Finkelstein 1988; Gal 1993; Zertal and Greenberg 1983; Zertal 1992). It is, however, recognized that settlement patterns cannot be meaningfully understood without investigating them within their geographical environment.[7] The method also establishes a link between a field survey and the excavation of a multi-period site (or sites) within the region (Portugali 1983; 1993; Kochavi 1989). The pottery sequences from excavations can thus serve as a check for the pottery found during the survey. This method, however, ignores the archaeology of the *total* landscape, and field systems never seem to be investigated. Attempts are made to understand settlement patterns and variations of land use within a total range by utilizing information on soils, rock types, vegetation and topography, which are usually extrapolated from existing maps. This is the closest, however, that this method comes to investigating landscape history.

Site catchment analysis (or site territory analysis) is the only method in which the landscape is seen as able to provide information of potential archaeological interest (Vita-Finzi and Higgs 1970; Higgs and Vita-Finzi 1972; 1975; Vita-Finzi 1978: 71-88).[8] Even with this method, however, the landscape is only trawled for information in so far as it can add to the understanding of the "site", i.e., the place of habitation. The method is not used without a site as its pivot and landscape history is not its aim. The information provided is extremely selective, and includes data on present-day soils, topography and vegetation. The method largely ignores changes to landscapes resulting from post-depositional processes, namely, soil erosion, colluviation and alluviation.[9] Other features of the landscape are either ignored or subsumed under the heading of "ethnographic data." A proper investigation of the archaeological features located in the landscape of a site's territory was never part of the method as originally formulated by Higgs and Vita-Finzi. As applied to ancient Israel/Palestine, this method is based on an idealized model of land use and fails to consider the complexity of human decision-making.

With conventional field survey procedures, it is sometimes difficult to draw a clear line separating the "site" (defined as the settlement, or the place of human occupation from its surrounding features (notably fields, roads, industrial installations and so forth). If the site is a city or town, then its fortifications could serve as a line of demarcation. However, what kind of separation does one make in the case of a farmstead? Does the "site" end at the outer boundary of the farm buildings, or does it also include the field systems and pastureland? What about the common land between the group of fields belonging to one farm and to another? If a given hill has scattered agricultural terraces and winepresses but no concentration of farm buildings, should it still be defined as a "site"?

With Landscape Archaeology, this problem simply does not exist because the "site" is actually the landscape and everything it contains, from settlements to fields. Nothing is excluded. In other words, this method does not concentrate exclusively on studying forms of settlement within a landscape. Instead, it aims at examining in a comprehensive fashion all the patterns of social activities evident within a given landscape, including those of settlements. This method moves away from the "sites in their setting" approach hitherto adopted by the "New" archaeologists working in the southern Levant since the 1970s (Dever 1988).[10]

The spatial limits chosen for the area of investigation in a Landscape Archaeology project, are entirely flexible and there are no steadfast rules. Thus, the scale of the area, whether large or small, depends on the goals set by the project director. The limits may be the boundaries of a specific geographical zone (for instance, the territory within a major valley system, a range of hills, or an extended plain) or of a more restricted microenvironment (for instance, a slope of a hill, a wadi system, or a small oasis) (Fig. 1-3).

In a Landscape Archaeology survey, all visible man-made features are identified and recorded, whether ancient or from the recent past.[11] Efforts are made to examine the varying associations between clusters of features within alternate geographical zones. The extent and function of all the dateable features in a landscape are examined and subsequently sorted chronologically. Many agricultural landscapes show evidence of the continuity of certain features through time (for instance, stone fences around fields, the topographical alignment of terraces, and the use of roads between plots of land). This is not surprising, because farmers cannot be oblivious to what preceded them in a landscape. Modern features should also be recorded because they may serve as the only existing record of antecedant features obliterated long before. For example, a row of trees planted along the edge of a field at the beginning of the 20th century may serve to delineate the original position of an ancient stone fence that at some point had been dismantled. While settlements undergo cycles of repair, rebuilding, extension and abandonment, many of the landscape features -- especially terraces, field walls and roads -- retain their original form and function and may continue to be used over thousands of years (Gibson 2001).

There are two major types of features normally recorded during a project: "upstanding features" (such as dolmens, stone fences, structures, lime kilns, and so forth), and "scattered features" (potsherds, flints, building debris, industrial wasters, and so forth). Once the features have been recorded during the survey and their spatial distribution sorted out on chronological maps, proper analysis commences. A key question to be asked regarding the recorded clusters of features is why the archaeological features are where they are. The additional questions posed depend on the nature of the landscape and the preservation of the remains. For example, to get from one feature to another which access routes (roads, paths) would need to have been used to get from feature to another? To what extent would rural settlements from the same period have been visible one from the other? Is it possible to link certain field systems with specific farms, villages or other forms of settlement? Is it possible to evaluate the quality of the land available for agricultural or pastoral purposes? What water sources are available? (Fig. 1-4) What sort of agricultural installations might be found adjacent to plots of land with certain types of soils? What areas were given over to industrial activities, such as stone quarrying, and charcoal and lime burning? Where were the cemeteries? And so forth.

Dating is always a major problem in a project based entirely on survey work. Hence, when possible, select key areas within a landscape should be tested by excavation to elucidate and clarify chronology especially in order to clarify the nature and extent of problematic buried features -- terraces, fences, stone clearance heaps, orchard planting holes, and so forth. This was possible during projects at Sataf and Modi'in. Such test excavations can serve as a useful "control" for the broader picture of the development of the landscape (Wilkinson 1991). Dating during a field survey may be possible by means of careful sherding, recognizing the morphology of features, or observing the stratigraphic and structural relationships between features. Discrete sherding should always be carried out in direct association with the appropriate feature being investigated (for example, in a plot of land surrounded by a stone fence, next to a wine press, along the course of a road, and so forth). A clear record should be made of the appearance and quantity of various types of sherd scatters that appear at different locations. These may appear as a random scatter (as in a field), an enclosed scatter (as in a cave), or a concentrated scatter (representing a focus of human activity, as around a cistern opening). Concentrations of sherds may be collected systematically within a grid laid out on the ground surface. Maps showing the quantified distribution of sherds, according to form and fabric, over defined parts of the landscape should also be attempted. If possible, well defined repeated sherding is worthwhile at single-period buried settlements, especially in areas being ploughed or with shifting sand dunes (such as along Israel's coastal plain) where old features disappear and new features appear. Another approach is to date features based on their morphology, which is sometimes possible for burial caves, agricultural installations, cisterns, and structures. Types of stone dressing and of plaster coatings used on the walls of caves and water cisterns may also be useful for dating purposes.

Finally, some features in the landscape may be dated on the basis of their stratigraphic proximity to other well-dated features. For example, the stone fences of fields abutting the edges of a Roman road must post-date the road itself (Figs. 1-5, 1-6). However, if these fences were partially destroyed by the construction of a Medieval fortress, then they can be dated to a point in time *between* the Roman and Medieval periods. Elsewhere in the same landscape, this system of fields might also be associated with farm buildings or with installations of a certain typological group of the Byzantine period, all of which combined together provide the supporting evidence for dating the development of the field system as a whole.

It is recommended that total pedestrian coverage be conducted of the area investigated in a Landscape Archaeology project. All features, or distinct clusters or units, should be numbered and described. Plans and sections of various upstanding features should be made. The field research may also be facilitated by geophysical prospection methods. Important tools include Remote Sensing Analysis (RSA), Global Positioning System (GPS) and Geographical Information Systems (GIS). The latter is particularly useful for land resources assessment (e.g. Burrough 1994, but there is an enormous new literature on the many possible applications of GIS that has now been published). Aerial photography is an extremely useful tool for Landscape Archaeology and can serve as the basis for field mapping and for identifying and plotting of features not visible at ground level (Kennedy 1989; Kennedy and Riley 1990; Cleave 1993; Kedar 1999; Kedar and Danin 2000). Topographical maps from the 19th century (such as the *Survey of Western Palestine*) and cadastral maps from the British Mandate period are also an important source of information (Gavish 1990; 1991). These can provide vital information on Arabic place-names, boundaries of agricultural lands, areas once covered with woodland, the location of cisterns, quarries, lime kilns, roads, paths, canals, religious buildings, and so forth.

Experience has shown that every region possesses a characteristic set of features that will inevitably be repeated in similar landscapes. For instance, features from an earlier period may have become useful depositories for stones cleared from the surface to create fields in later periods. Similarly, winepresses would be cut in areas of rocky outcrops within agricultural areas, burial caves hewn into rock scarps with a natural step-like contour, and so forth. During a survey, notes should be made on factors affecting archaeological visibility: vegetation cover, alluvial deposits, hill-slope erosion and modern development.

Collaborative work with a geomorphologist can be extremely illuminating, especially since the deep alluvium in parts of Israel has resulted in the burial of large tracts of ancient land surfaces and early settlements that can be undetectable without excavation.

Examples of Projects of Landscape Archaeology

The following are summaries of the results obtained during four Landscape Archaeology projects conducted in Israel since 1987.

Sataf

Sataf is located in the hills to the west of Jerusalem and consists of a steep slope covered with terraces overlooking the Soreq Valley, with two springs of water (Ein Sataf and Ein Bikura) and the ruins of a village abandoned in 1948 (Fig. 1-7). It was targeted as an ideal location for a project of Landscape Archaeology precisely because it was *not* regarded as an archaeological site (the only ancient remains found previously were stray prehistoric flints and an isolated Roman tomb). The goals of the project were to survey the landscape, to investigate the spring systems, to excavate various features and to cut sections through some of the agricultural terraces. Work commenced in 1987 and was completed in 1989 (Gibson et al. 1991).

A total area of 90 dunams was surveyed. A map of all man-made features visible in the landscape was prepared on the basis of aerial photographs, and it was subsequently checked and corrected in the field. A detailed description was made of all visible features, ancient and modern. These included structures, terraces, paths, installations, cupmarks, and caves. A detailed study was made of the springs and their component parts (caves, flow-tunnels and pools), including recording the layers of plaster on the pool walls. A detailed record was made of the complex system of irrigation channels extending along the terraces below the springs.

Trial excavations were conducted at 18 separate locations and three larger areas were later excavated to clarify the stratigraphy of settlement remains previously uncovered in the trial areas. The history of the Sataf landscape was traced from the Chalcolithic period and to the present.

The remains of a Chalcolithic settlement (late 4th millennium BC) were uncovered on rocky terraces above one of the springs. This settlement consisting of houses and caves, was later buried beneath agricultural terraces dating to the Ottoman period and none of the remains were visible prior to the excavations. It was probably a small settlement (estimated size: 5.4 dunam). Radiocarbon dates were obtained from carbonized short-life materials taken from the floors of excavated structures. Woodland clearance appears to have taken place in the area close to the settlement and near the spring. The inhabitants practiced dry farming, with the cultivation of some grain crops (wheat and possibly barley), and animal husbandry (with sheep/goats predominating, and some cattle). The olives (*Olea europa*) may have been gathered wild. Nothing is known regarding the location and appearance of the fields from this period.

Early Bronze Age I remains (early 3rd millennium BC) were found scattered at a number of different locations in the Sataf landscape, buried beneath later agricultural terraces. The location of structures at distances of as much as a quarter of a kilometer from each other has important implications regarding the landscape at that time. First, it indicates that the EBI village was not nucleated, but consisted of scattered dwellings, each structure probably with its own fields and animal pens. It is highly unlikely that the entire slope was densely covered with houses at this time. The dispersed remains from Sataf may also help to explain why so few EBI sites have been found during surveys in the terraced highland zones. Nucleated remains are much more likely to be detected during a survey than dispersed settlements, especially if they are hidden beneath later terraces. Second, the dispersed EBI remains imply that during this period, substantial woodland clearance must have occurred in the area of the two springs and in the surrounding landscape. This would inevitably have led to concurrent terrace construction: one example of an EBI terrace was found in Area B (Fig. 1-8). The surviving plant remains from the period indicate that there was a shift towards the cultivation of the vine, and perhaps also of olives, with a decline in the cultivation of grain crops. The decline in the use of flint may also hint at the increased use of copper implements for agricultural work. This accords with the current view that there was a perceptible shift from small groups practising subsistence agriculture and pastoralism to large communities specializing in horticulture in the highlands during the transition between the Chalcolithic and Early Bronze Age (Finkelstein and Gophna 1993).

Worn sherds were found dating to the Middle Bronze Age II and Iron Age II periods, and the evidence indicates that substantial agricultural activities were renewed in the Sataf landscape only during the late Hellenistic period (from the late 2nd century BC) and continued throughout the Early Roman period until the end of the 1st century AD. It appears that the various components of the spring system of Ein Bikura were first constructed during the Early Roman period. The source of the spring was opened up by enlarging the cave and cutting channels in the bedrock to divert water into a large collecting pool. A trench excavated next to the original wall of the pool confirmed that it had originally been built in the Early Roman period. There is no evidence, however, that the flow-tunnel was in use at this time, and the area between the cave and the pool may have remained open. Irrigation terraces would have been built near these structures and under the level of the pool of the spring, but could not be excavated. Terraces used for dry farming were uncovered elsewhere on the slope above the spring. The remains of a well-preserved farm building surrounded by terraces for dry farming were investigated along the southwestern limits of the agricultural lands of Sataf. A contemporary Jewish stepped ritual bath (*miqweh*), with walls coated with plaster, was found in the vicinity of the structure. Such installations were frequently located on farms and were used for ritual cleansing between agricultural activities (cf. Mishna *Tohoroth* 10:3; Danby ed.).

The Sataf landscape was modified and expanded during the Byzantine and Early Islamic periods (4th-8th centuries AD), with the rebuilding of the water systems and the construction of more widespread terracing on the slopes of the surrounding hills. The water systems consisted of flow-tunnels and large pools, irrigation channels and at least three small water collection pools scattered in the terraced area. The last imply a certain degree of land-use intensification. The flow tunnel and large pool of the spring of Ein Sataf were apparently first built during this period (Fig. 1-9). The older water system at Ein Bikura was reconstructed and a flow-tunnel added. The lower part of the pool was converted into a fishpond, with rows of ceramic jars built into the walls apparently for breeding of fish (similar installations are known from the Byzantine period, mostly on the coastal plain). Substantial terracing activities also took place at this time. A large farming complex was surveyed at nearby esh-Sheikh 'Ubeid. Agricultural terraces were contiguous with this building and there were winepresses (one which had a screw-press device) nearby.

No archaeological evidence has been found attesting to activities at Sataf from c. 750 AD to the Mamluke period, despite the proximity of a 12th-century castle (*Belmont*) at nearby Suba (Harper and Pringle 2000). The two water systems were reconstructed during the Mamluke period. The barrel-vaulted room next to the pool of Ein Sataf and the eastern extension of the flow tunnel of Ein Bikura are dated to this period. The excavation of a trench in the cave of Ein Sataf revealed a stratified sequence of deposits. The earliest phase was represented by a floor with Mamluke material sealing a reconstruction of the flow-tunnel that ran from the cave to the pool outside. On top of this floor was a fall of rock that had resulted from a substantial collapse of the cave ceiling, perhaps due to an earthquake. Above it was a floor with hearths, pottery and three coins, two from the floor matrix (AD 1363-77) and one from the floor surface (AD 1382). Sataf and its agricultural lands are known to have been part of a religious endowment made in 1320 by Shu'ayb Abu Madyan to the Maghribi community for the purpose of the construction of a hostel next to the Temple Mount in

Jerusalem. Therefore, the water systems at Sataf were probably reconstructed round about AD 1320.

The entire landscape of Sataf was remodified in the course of the Ottoman period, with the restoration of the spring systems and the construction of numerous terraces. Sataf appears in the early Ottoman tax registers of 1525-1526 as a *mazra'ah* -- an uninhabited place with agricultural land belonging to one of the neighbouring villages. However, the tax registers from the years 1538-1539 list Sataf as both a *qaryah* (village) and as a *mazra'ah*. This confusion may have arisen from the fact that during the early Ottoman period, some farmers appear to have lived in cave dwellings at the site but the rest lived in a neighbouring village, probably Bahtiyar. Several cave dwellings were investigated at the site and these could not strictly be described as representing a "village." Sataf was not mentioned in the tax registers for 1596-1597, but the adjoining village of Bahtiyar was. A process of settlement nucleation appears to have taken place at Sataf in the course of the 18th and 19th centuries, perhaps as a response to inter-village conflict. It remained one of the smallest villages in the Beni Hasan District of the Jerusalem *sanjaq*. During the early 20th century, the settlement dispersed in the landscape and terraces were repaired. The settlement and its terraces continued to survive until 1948. Terraces have been restored as part of a project in which the site has been converted into a recreational park.

Two important conclusions emerged as a result of the work at Sataf. First, the discovery of stratified protohistoric remains hidden beneath agricultural terraces was unexpected. The implication is that many more ancient settlements in the highlands of Palestine may have similarly "disappeared" under later agricultural terracing. This conclusion will undoubtedly affect our present understanding of settlement patterns in the central highlands during the protohistoric and early historic periods (Gibson 2001). Second, the excavation of the agricultural terraces at Sataf has shown this to be a worthwhile and productive archaeological pursuit. Such an investigation can reveal the history of individual terraces and the development of a terrace system as a whole.

The Golan Heights

Between 1983 and 1988, a detailed examination was made of the landscapes surrounding four villages in the central Golan (Kafr Naffakh, Na`aran, Farj and er-Ramthaniyyeh). This landscape archaeology project was intended to complement the detailed topographic and architectural survey of the settlements conducted by Claudine Dauphin from 1978-1988 (Gibson and Dauphin 1990; Dauphin and Gibson 1992-93).

The investigation of the field systems was conducted on foot, using maps based on aerial photographs to record field walls, stone heaps, dolmens, ruined structures, animal pens, enclosures, and other archaeological features, as well as roads and paths. A detailed catalogue of the features was prepared. The investigation of the field systems was particularly rewarding since the landscapes of the Golan have been left relatively untouched by modern development. Our investigations allowed us to distinguish between various types of ancient and modern field-wall constructions. For instance, ancient field walls were frequently built of boulders, whereas small rounded fieldstones were usually used for their restoration in the 19th and 20th centuries AD and to build new walls. Moreover, the relationship between ancient houses and contiguous field wall systems on the outskirts of settlements was studied at all sites. An attempt was made to locate farmsteads and other rural structures, such as field towers and wine and oil presses, in the vicinity of the villages. Parts of the landscape given over to pasturage or industrial activity were also investigated. Ancient road and path networks and natural topographical features enabled us to determine the extent of the lands belonging to each separate settlement. Potsherd scatters associated with a given feature or field unit were examined.

Kafr Naffakh is a compact densely built-up village. Many of the houses are of Syrian date and were built directly over Byzantine structures. One of the courtyards had the remains of an oil press. An ancient paved road passed the settlement to the south. As a result of modern military activities, little of the agricultural system surrounding the settlement survived, except in the northern parts of the wadi to the southeast. The system consisted of a dam-wall built of boulders, 1.8 m high, retaining an expanse of soil on its northern side. Winter rains were ponded behind the dam-wall for the purposes of cultivation and then filtered through the soil and the wall to the wadi bed below.

Na'aran is located on the same ancient route as Kafr Naffakh. The settlement has an upper area built on a volcanic hill and a lower area, where the remains of a bathhouse dating from the Roman period were found. The settlement was resettled in the Byzantine period, as attested by a carved inscription from the 6th century AD. A rock-cut aqueduct brought water, probably from the spring of 'Ein el-Ayadeh, to the settlement. A seasonal spring cave was examined in a rocky scarp on the eastern bank of the wadi to the east of the main settlement. Field systems were investigated to the north, east and south of Na'aran. A close examination was made of 67 individual plots of land, their internal features and boundary walls, and sample areas were sherded. Three stages of human activity were detected within the landscape. Stage I is represented by dolmen fields. There are at least three types: boulder dolmens with an outer ring of stones and a central built chamber lined with vertical stones; large heaps of small- or medium-sized fieldstones; and simple cist graves, which are either solid-built or lined with vertical stones. The dolmens are scattered across the landscape, seemingly unassociated with a settlement. Bronze Age sherds were occasionally found on the field surfaces. Stage II consists of agricultural fields. To the southeast of the settlement, earthen embankments are gently stepped from north to south with enclosure walls built of large boulders and fieldstones. Boulders taken from dismantled dolmens were sometimes reused in boundary walls. Smaller plots on the northern side of the system, close to the road leading into the settlement from the east, may have been used for vineyards in the Roman and Byzantine periods. Surface sherds from the fields are indicative of manuring practices. In Stage III, the field enclosure walls at some distance from the settlement were reused from the preceding stage of construction, while those in close proximity were restored with small fieldstones. This stage probably represents the Turkish to Syrian occupation of Na'aran up to 1967.

Farj consists of an accumulation of ancient structures around three pools. Structures dating from the Roman and Byzantine periods were investigated, some of which were restored in the Mamluke period. The site was resettled by the Circassians in the second half of the 19th century. The field systems were investigated within a radius of 1.5 km and to the north as far as the village of el-Fakhm. Two hundred plots of land were closely examined and sample areas were sherded. The landscape underwent four stages of development. In Stage I, large numbers of dolmens extended over the entire area. Dolmen fields had not previously been recorded for the Farj region. A sample of 50 dolmens was examined. Dolmen clusters tended to be concentrated in stony areas unsuitable for cultivation. Bronze Age sherds were collected in the fields. Stage II is represented by rectangular fields with boundary walls and earthen banks associated with the Late Hellenistic and Early Roman quarters to the northeast and southeast of the settlement. Stage III is represented by a well planned and regulated complex of field systems divided into strips with straight boundary walls (Fig. 1-10). It was situated east of the settlement and had a northwest to southeast axis, extending for a distance of 1 km. To the northeast, there were stone-cleared areas suitable for grazing. Areas located at a distance of 500 meters to the east of the settlement were comprised of regulated clearances with boundary walls. This stage should probably be associated with the Late Roman to Byzantine occupation at Farj. Stage IV includes the modern reconstruction of field systems close to the settlement or alongside roads leading out of the settlement. Curvilinear animal pens are located on the periphery of the settlement to the north and south.

Er-Ramthaniyye is a densely built up settlement from the Roman and Byzantine periods located on the rocky summit and slopes of a volcanic plug, and part of the plain below it (Fig. 1-11). The site is well known in the history of Jewish settlement, for the failed attempt to establish a Benei Yehuda colony there between 1885-1887. Eleven sources of water were identified in and around the settlement. A protohistoric settlement was found 500 metres to the southeast of the hill of er-Ramthaniyye, covering an area of 1.7 hectare. Bronze Age pottery was collected in the course of the sherding of agricultural plots up to a distance of 1 km away from the settlement, suggesting woodland clearance and an intensive agricultural use of the land even at this early date. The megalithic landscape of er-Ramthaniyye consists of a field of 86 dolmens, 56 of which are located in the rocky zone to the southeast of the hill. All the megaliths, with the exception of one free-standing type of dolmen, consist of burial cists within tumuli. Previous studies of dolmens have concentrated on their typology, function and date, and little work has been done on the spatial distribution of major types of dolmens within the landscapes of the Golan. The agricultural lands of er-Ramthaniyye extended over an area of approximately 6 square kilometers and 222 plots of land were examined within a radius of one kilometre around the hill of er-Ramthaniyye. The ancient field system consisted of rectangular plots. As opposed to the shape of individual plots of land, the blocks of fields were shaped in conformity with natural topographical features and the position of pre-existing local roads. These roads radiate out from the settlement in all directions, linking it with villages in the vicinity. Large quantities of Roman pottery were found scattered over the entire area of the field systems. The pottery was probably deposited in the fields as one of various soil-improving agents including manure, domestic rubbish and so forth. In the Byzantine period, er-Ramthaniyye became the focus of Ghassanid Christian nomads, who flocked to the site to commemorate the memory of St. John the Baptist. Their encampments, or *hirtat*, consisted of clusters of curvilinear enclosures serving as tent bases, which were found during the survey mainly to the east of the settlement. The largest of these covers an area of 200 x 250 meters. (Fig. 1-12) Also from this period are elements of a ritual landscape. Liturgical processions appear to have linked the major roads leading from the north, south and east, towards the large pool of water at the eastern foot of the hill of er-Ramthaniyye. Encampments were positioned along these routes in association with outdoor altars, standing stones -- some of which were carved with grooves and crosses – and unique petroglyphs that have been identified as maps, showing the site, the sources of water and the layout of roads in the surrounding landscape. Because of the simplicity of the actual landscape these maps undoubtedly would have served a ceremonial rather than a functional purpose. [12]

Dor

The project was conducted in the autumn of 1994 in an area of eight one-square kilometres within the hinterland of Tel Dor (Tanturah) on the northern coastal plain (Gibson et al. 1999).

Following a number of preparatory steps, a map of the region was drawn based on an aerial photograph (Fig. 1-13). This master map depicted all visible artificial man-made features, whether ancient or modern, and the boundaries of respective geographical units. Copies of the map were subsequently annotated and corrected in the course of the field survey. All archaeological activities visible, irrespective of surface configurations, were designated "features" and numbered accordingly. Twenty-two classes of features were found in the area, including sherd scatters, piles of ashlars, stone fences, caves, cist graves, cisterns, reservoirs, wine and oil presses, *columbaria*, quarries and so forth. In total, 447 features were examined during the survey. Ultimately, the overall analysis of these features was accomplished by means of the careful examination of inter-landscape clustering effects within alternate geographical zones. The distinctive regional patterning of these "clusters" of features was then combined with additional data derived from morphology characteristics and surface ceramic finds. There were five survey zones: Zone I included the offshore islets, coastal beaches and backshore fringe; Zone II the western *kurkar* ridge; Zone III the western alluvial plain; Zone IV the eastern *kurkar* ridge; and Zone V the eastern alluvial plain extending up to the Lower Carmel Hills.

Methods of fieldwalking differed substantially from one sub-unit of the landscape to another. In flat areas of alluvium, fieldwalking was only possible in lands that were not under cultivation at the time of the survey. These areas consisted of unploughed, ploughed and deep-ploughed lands. The size and shape of each plot of land was marked on a map and given a feature number. Fieldwalking was undertaken along the regular grooves created by tractor-ploughing activities. An average distance of 3 metres was maintained between one investigator and another. Artifacts collected included potsherds, flints, basalt and marble objects, stone tesserae, bones and shells, all of which were quantified. Along the *kurkar* ridges covered with dense thorn bushes, a different method of fieldwalking was used. Sherding was undertaken next to each specific feature under investigation which was numbered accordingly. These features were comprised of a broad range of archaeological remains, including complete small settlements, farmhouses, stone quarries, agricultural installations (mainly oil and wine presses), caves, tombs, wells, cisterns, and so forth. A total of 20,365 artifacts were collected from 447 features in the course of the six-week survey, with more than 550 artifacts processed daily.

Apart from the multi-period Tel Dor and the Ottoman town of Tanturah, all ancient settlement remains found during the survey were situated on the elevated ridges of Zones II and IV.

Open-air stations, probably fairly temporary in character, were already being established along the *kurkar* ridges by the Neolithic period. A large cluster of features with building remains and pottery and flint scatters dating from the late Neolithic (Wadi Rabah), Chalcolithic and Early Bronze I periods was investigated south of Moshav Dor in Zone II. Remains of a settlement and of a tower (?) dating from the Middle Bronze Age IIA were investigated at two locations along the *kurkar* ridge in Zone IV. The size and character of the settlement fits in well with the overall picture of small unwalled MBIIA villages and hamlets scattered along the *kurkar* ridges in the northern coastal plain. The survey evidence also points to extremely limited agricultural and industrial activities in the countryside in the course of the Iron Age and the early part of the Persian period (sixth-fifth centuries BC). The extremely low densities of Persian pottery throughout the survey region suggest that until as late as the second century BC, occupation was restricted solely to Tel Dor and its immediate surrounding fringe (i.e. the lower city to its north and east). It is possible that these extra-mural suburbs were of an agricultural character and that the first attempt to drain the surrounding swamps was being made at this time. The drainage of swamps would have continued in the course of the Hellenistic period, a process that based on the archaeological evidence, seems to have been largely accomplished by the late second century BC. The remaining archaeological features investigated in the landscape, of an agricultural, industrial and sepulchral character, are all associated with varying concentrations and quantities of pottery dating from the Hellenistic through the late Ottoman periods. A good number of farms or hamlets were investigated that appear to have been founded in the Roman period, many of which continued in use into the Byzantine period. Eleven clusters of features were examined and 4,729 sherds were collected, with 88.2% of this material of Byzantine date. The early Islamic period (Umayyad and Abbasid) was not represented by features in the landscape, except for a few isolated sherds. The latest period that is well represented in the landscape is the Ottoman period, from the 16th or 17th centuries through the early part of the 20th century. It would appear that throughout Zone IV, a substantial clearance was made of building stones and ashlars from earlier Roman and Byzantine settlements, presumably for reuse in the construction of the houses of the town of Tanturah in Zone I. Late Ottoman period Bedouin encampments and cemeteries were also investigated.

Modi'in

From 1995-1999, a project of Landscape Archaeology and salvage excavation was conducted in the hills of Modi'in by the author, in collaboration with Egon Lass on behalf of the Israel Antiquities Authority (Gibson and Lass 2000).

Modi'in is located in the northern Shephelah, in the western foothills, not far away from the market town of Lod. The project was initiated as a result of the immediate need for a series of salvage excavations within the area of the new city of Modi'in, which was then under construction. Historically, Modi'in was the hometown of the Maccabees who rebelled against the Seleucids in the mid-second century BC. The author believes that the ancient town of Modi'in should be identified with Khirbet el-Burj (Titura) and not, as traditionally held, with Khirbet el-Midye, which is essentially a Byzantine settlement and cemetery.

The Modi'in region is dominated by low hills and has an undulating and rugged appearance, with enormous expanses of rocky outcrops, great quantities of loose stones and small pockets of terra rossa soil. Since 1995, at least 2000 archaeological features have been recorded and of these a selection was chosen for excavation. The choice of the areas sampled was not determined on scientific grounds but decided by the Ministry of Housing and Construction and individual contractors. These salvage excavations are still in progress as the modern city continues to expand. The archaeological remains of ancient human activities were all designated "features" and numbered accordingly. The following classes of features were encountered in the survey and excavations: farm buildings, towers, cisterns, sherd scatters, prehistoric flint scatters, roads, terraces, stone fences, stone clearance heaps, threshing floors, caves, tombs, wine presses, cupmarks, stone quarries, lime kilns, and charcoal burners. (Fig. 1-14) These features are located in units of land belonging to the three main ancient settlements in the region, namely, Khirbet el-Burj (Titura), Bir Ma'in (Re'ut) and Berfilya.

Excluding the farmhouses, which had an abundance of pottery and other artifacts, the material culture from the features was generally sparse, with no more than a handful of potsherds per feature. The pottery from the survey and excavations may be attributed to five main chronological groups: (1) Chalcolithic to Middle Bronze II; (2) Iron Age II; (3) Late Hellenistic to Roman; (4) Byzantine to early Islamic; (5) Medieval to Ottoman. The fabrics of the five pottery groups were sufficiently distinctive to allow for the dating of abraded body sherds even when components such as rims, handles and bases were not available, using the material derived from the settlements (stratified vessel types) as a "control." The range of pottery types used by the ancient peoples working in this landscape was limited -- mainly storage and cooking wares -- and this also helped with the pottery identification procedures. A pottery typology was eventually created based on the material derived from the various excavations.

Methods of excavation varied from one feature to another. In dealing with a stone clearance heap, for example, it was essential to distinguish between initial stone clearance activities resulting from field or terrace construction and those resulting from field or terrace ploughing. (Fig. 1-15) The dating evidence available from a specific feature could then be interlinked with the patterning of other dating materials derived from clusters of features across the entire landscape. There seemed to be little point in excavating features as mere dots within the landscape without investigating the overall context of such features.

The excavation of farm buildings helped to provide more detailed information regarding ceramic sequences that could then be used to date nearby features in which only very few dating materials had been preserved. (Fig. 1-16) There was a great difficulty in dating rock-cut winepresses and cupmarks, many of which were visibly exposed without much soil within them. They were generally dated to the late Hellenistic-Byzantine periods on morphological grounds, as well as on the basis of the assumption that the sherd scatters found in their immediate proximity dated from the time they were in use. In the case of lime kilns and charcoal burners, it became evident that they could not be dated properly without radiocarbon dating of the charcoal deposits on their floors.

The best soils in the Modi'in region were in the shallow but narrow wadis and in flat catchment areas of soil. The slopes of hills were largely rocky, without much soil. Some of the pockets of soil were so small that it seemed hardly likely that they would have been utilized for cultivation purposes, but excavations showed that these pockets were indeed used for agriculture and not just for hoe cultivation. Plough-marks were clearly evident on the surfaces of the underlying rock beneath the pockets of soil and on the edges of boulders adjoining the plots.

The region was criss-crossed by a number of ancient highways and regional roads linking the principal settlements with the market towns of the northern Shephelah. The Roman highways have "bends" or "shoulders" at regular intervals of about one kilometer, characteristic of roads of this period in the Levant (as shown by the late Derrick Riley). The regional roads were adapted to the topography of the area and the layout of the field systems, giving farmers easy access to their fields from their place of abode.

A great deal of effort was expended in antiquity on the layout of field systems in the area, with plots of land surrounded by stone boundary walls and terracing on the slopes. A sample of terraces, wadi dam walls and field boundary walls were excavated. The construction techniques of these terraces and dams resemble those investigated in other highland environments, especially in the use of stony drainage fills behind external retaining walls. The boundary walls of the fields and the terrace walls were originally built on a general north-south axis, which was evident in many areas of the Modi'in landscape. This co-axial arrangement indicates that field systems were pre-planned and not merely the result of a general adaptation to the existing topography and environmental conditions. The earliest field systems in the Modi'in region may be dated to the Hellenistic period. A massive stone clearance operation must have been undertaken in the area immediately prior to the construction of the fields. It is not surprising therefore that thousands of stone clearance piles (*rujum* in Arabic) were recorded in the Modi'in area, ranging in size between a few meters to as many as eight metres in diameter. They usually had ring walls built of boulders, except for the very small examples. The fact that these piles were sometimes in use over long periods of time made it possible to study their internal stratigraphy.

The survey and excavations conducted in the Modi'in region indicate two main periods of extensive agricultural exploitation in the landscape, the first from the late Hellenistic-early Roman period (late second century BC to the first century AD) and the second from the Byzantine-early Islamic period (sixth to ninth centuries AD). Not surprisingly, the only two farmsteads excavated in the region were also in use during these two main periods. Prior to the Hellenistic period, agriculture was much more intensive and was concentrated in the wadi beds. In the Medieval and Ottoman periods, the region was almost exclusively utilized for industrial purposes, with the establishment of hundreds of lime kilns and charcoal burners across the landscape (Fig. 1-17).

Conclusions

Landscape Archaeology has only gradually been adopted as a genuine archaeological research tool in the Near East in general and in the Levant in particular. One of the reasons for this has been the dominance of Settlement Archaeology, and another that scientific work on landscape history has sometimes been regarded as the privileged domain of geographers and cultural historians.

Landscape Archaeology clearly does not need to compete with conventional field survey methodologies. Indeed, its methods are sufficiently flexible and can easily be combined with existing regional surveys or excavation projects. Moreover, excavations play an important part in the investigation of key features within a landscape, especially for dating purposes. Landscape Archaeology provides data on land exploitation systems that were previously unobtainable using conventional methods of field survey or excavation. For example, the study of ancient field systems in the Levant was totally ignored by conventional archaeology, but has developed largely through the use of Landscape Archaeology methods.

The object of Landscape Archaeology is to examine the extent to which landscapes have been modified by human agencies from antiquity to the present day, to trace the intricate changes and transformations from wildscape to landscape as we now know it. This knowledge can help to articulate the very complex history of economic systems and social transformations in the Levant. It is hoped that the scientific inquiry into the history of landscapes will eventually become an integral part of all Near Eastern archaeological programs, and those in the Levant in particular.

Notes

1 This article is based on material that first appeared in my unpublished doctoral thesis entitled "Landscape Archaeology and Ancient Agricultural Systems in Palestine" which was submitted in 1995 to the Institute of Archaeology, University College (London). My supervisor was Peter Parr and I am very grateful to him for his unstinting support over the years, as well as to Gordon Hillman, Ken Thomas and many other friends and colleagues. I would like to acknowledge receipt of two major grants that facilitated my PhD research, namely a Major State Studentship from the British Academy (1986-89) and a Wingate Scholarship (1989-91). During the 1990s I was a post-doctoral fellow at the W. F. Albright Institute of Archaeological Research (Jerusalem); my thanks to the Director, Sy Gitin, for his encouragement. I would like to extend my thanks to Edna Sachar for her excellent comments on an earlier version of this article.

2 The problem of defining an archaeological "site" has given rise to much debate over the years (notably Keller and Rupp 1983; Macready and Thompson 1985; Wagstaff 1987; Fiches and Van Der Leeuw 1990; Barker and Lloyd 1991; Schofield 1991). Numerous arguments have been made as to what constitutes a "site" and what does not. Attempts have been made by some to delimit "on-site" and "off-site" features. Others have used the word "site" to describe the remains of only *significant* human activity and "findspot" for everything else (Lloyd et al. 1985). There are also those who make a distinction between concentrated and dispersed scatters of artifacts, which are primarily potsherds. Problems do arise, however, when one field worker refers to a few scattered potsherds merely as "background noise" (or non-settlement evidence, cf. Cherry 1983), while another will describe the same handful of potsherds as a fully-fledged "site." A reasonable assumption is that while potsherds do tend to peak in intensity close to settlements or at locations of prolonged human activity, there will be certain other locations (e.g. stone quarries) which are totally devoid of artifacts that still deserve to be described as "sites." The exact definition of a "site" is only of importance if the primary focus of the research is on settlement patterns, otherwise settlements are to be regarded as part of the overall mosaic of "features" that make up a given landscape.

3 Cherry (1983: 390) has written that "it is quite impossible to lay down golden rules or universally applicable procedures, cookbook style" on such matters.

4 In some parts of the country, notably the semi-arid and desert regions, the separation between "sites" may be defined by an absence of features in the empty or blank areas between concentrated remains of human activity, regardless of whether these are of nomadic camping sites or fortresses. However, potsherd and lithic scatters at locations without obvious architectural remains may be difficult to detect without the optimum total pedestrian coverage of the survey area. Walking transects in specific directions at prescribed distances, or sampling a proportion of the area using a statistically acceptable technique, may significantly increase the number of features found as opposed to undertaking random surveys (MacDonald 1988; Rosen 1987).

5 Since the well-known *Survey of Western Palestine* conducted in the 1870s (Conder and Kitchener 1880; 1882; 1883), no other systematic countrywide survey was carried out in the region until the Archaeological Survey of Palestine which was begun by P.L.O. Guy in the years 1937-39 but was later discontinued (Gibson 1999: 123). Following the establishment of the Israel Department of Antiquities in July 1948, it was decided, perhaps at the instigation of one of its founding members, P. L. O. Guy, that a complete archaeological survey was needed within the territory of the new State of Israel. However, it was not until July 1962 that the Israel Archaeological Council

met and decided that a new survey should be done. Consequently, in July 1964 the "Society of the Archaeological Survey of Israel" (ASI) was founded with the participation of many of the archaeological institutions in Israel. Numerous surveys have been conducted since then but only a small percentage of these have been published.

6 It is surprising that the Archaeological Survey of Israel never issued guidelines to its surveyors with information about survey methods, on the use of consistent descriptive terms, on the minimum required measurements needed when recording visible ancient features (particularly caves and upstanding structures), on the techniques of sherding sites, and so forth. As a result, ASI surveys have been carried out to widely differing standards (Rosen 1987: 31) and a comparative study of the results is nigh virtually impossible. Unfortunately, the ASI publications rarely include information about the methods used by the surveyors; the exceptions are Haiman (1986) and Rosen (1987; 1992).

7 The idea behind this approach is that human behaviour tends to be regionally circumscribed in a balanced fashion (cf. Binford 1964). Many of the regional archaeological projects are therefore based on methods adapted from studies in human geography (Wagstaff 1987). Regional surveys can generate a large amount of data regarding settlements, their sizes and distribution. However, the method makes no provision for the fact that the visibility of "sites" may sometimes be culture-specific. This could give rise to the mistaken interpretation, for example, that *visible* sites must represent settlements that are only of a sedentary status, whereas *invisible* sites (i.e. sites not visible to the fieldworker) must reflect nomadization.

8 The idea behind site catchment analysis (SCA) is that only a partial picture of the economy of a site may be provided by the study of the faunal and botanical remains from excavations, and that this information could be substantially expanded by the study of the economic viability of the site's territory and this is done by examining its present-day location, topography, soils and vegetation. The "site", or the place showing evidence of human activity, serves as the starting point for the research on the surrounding territory or its "catchment". Chisholm in his study of traditional agricultural subsistence communities had put forward the theory of diminishing returns in direct relation to the distance from a site (i.e. economic distance) and this became one of the adopted principals of SCA (Chisholm 1968, 43-66). In other words, lands cultivated beyond a walking distance of 1 km from the site showed a decline in net return and even more so at a distance of 3-4 km. Hence, the working model used by the SCA method was that lands located outside the 5 km radius from the site were very rarely cultivated and this was because at this distance the costs would always outweigh the benefits. In order to resolve the difficulty of the effect of topography upon linear distance, radii were established using time distances of one-hour walks. This method was improved by Bailey and Davidson (1983) who suggested using isochronic distance: the average time needed to walk 5 km on flat land is one hour but an extra half an hour should be added for every elevation of 300 m. Wagstaff commented that this improved method "has the advantages of allowing for topographical variation on the configuration of site territories at all points and, being easy to apply to the analysis of topographical maps, of removing the need for much time-consuming field work and more or less extensive interpolation." (Wagstaff 1988, 66). Since the method was devised, projects ultilising SCA have been undertaken in different parts of the world with varying degrees of success (Webley 1972; Foley 1977; Roper 1979; Tiffany and Abbot 1982; Dennell 1987; Wagstaff 1988).

9 The composition of soils, their texture, depth and drainage, may be changed quite drastically over time by human modifications (Bell and Boardman 1992). In fact, soils can change even over relatively short periods of time, as Rosen (1986) showed in her study of environmental change during historic periods around Tel Lachish. Hence, Portugali's reconstruction of land use in the environs of the Iron Age settlement at Tel Qiri in the Jezreel Valley, using a modified version of SCA, would seem to be fundamentally flawed (Portugali 1987).

10 My definition of landscape archaeology as a form of "total archaeology" differs from the one proposed by Wilkinson (1991) in which he says that it deals with the "features spread *between* (my italics: S.G.) conventional occupation sites, as well as the smaller sites usually passed over by most surveys."

11 Banning (1986), who has had extensive experience working in Jordan, is opposed to landscape archaeology on the grounds that it is impractical because of financial restraints and the limited time available for surveying. He wrote: "it is unrealistic to suppose that anyone really records as sites all spatial loci that exhibit evidence of human behaviour or activity. Apart from the difficulty in distinguishing primary from secondary deposits of cultural material, we cannot maintain that anyone records every isolated artifact, pit, fence or rock-pile, ancient or modern, in the site inventory. The volume of data in a country like Jordan would be overwhelming, and compromises are inevitable. Most researchers actually record only tells, settlement sites, loci that exhibit traces of architecture, and major cemeteries. This is consistent with an interest in ceramic chronology and 'urban' architecture and to that extent the practice is quite justified..." (1986: 26). Banning's views on the merits of landscape archaeology appear to be unnecessarily dismissive. Surely the difficulties inherent in the study of ancient exploitation systems within landscapes should be regarded as challenges to be overcome (cf. Vallat 1991)? In any case, one method of survey does not necessarily have to be exclusive of the other. While methods of landscape archaeology are admittedly difficult to carry out on a regional scale, this does not mean that

micro-regions cannot be sampled. Indeed, only landscape archaeology can provide data on the complexity of human decision-making in manipulating, modifying and utilizing land surfaces. Critical socio-economic issues such as patterns of land ownership and systems of farming will only emerge if every aspect of the landscape is sampled. Barker has pointed out that "survey, like excavation, necessitates choices about goals and techniques; theoretically, we may choose one survey method to look for mountains and another for molehills. At the same time, however, the onus is on the mountain surveyor to demonstrate satisfactorily either that molehills did not exist or, if they did (as seems normally to be the case), that their deliberate omission was regarded as justifiable given the stated research goals of the project" (Barker and Lloyd 1991: 3). While Banning, the "mountain surveyor", must surely be right in his call for systematic sampling procedures to be used during regional surveys of settlements, at the same time he is quite wrong in ignoring the possibilities inherent in sampling micro-regional features, the "molehills", as well.

12 A number of stones carved with grooved lines and cupmarks – known as petroglyphs - have been identified as possible maps. Some of these date from as early as the Bronze Age but there are also examples from the Byzantine period. A petroglyph dated to 5,000 years ago was discovered in 1978 in a cave in the Jebel Amud, to the southeast of Petra. According to Borzatti (1994) the stone (5 metres in diameter) has a series of markings on it which are interpreted by him as representing a map of the Bronze Age villages in the region. Similar petroglyphs were investigated as early as the 1880s by Conder during his Survey of Eastern Palestine. One such "map" consisted of cupmarks and lines carved into the upper surface of a stone dolmen to the south of Amman (Conder 1889). Its date is uncertain.

REFERENCES

Allen, M.J.
1991 Analysing the Landscape: A Geographical Approach to Archaeological Problems. Pp. 39-57 in *Interpreting Artefact Scatters: Contributions to Ploughzone Archaeology,* ed. A. J. Schofield. Oxford: Oxbow.

Aston, M.
1985 *Interpreting the Landscape: Landscape Archaeology in Local Studies*. London: Batsford.

Aston, M., and Rowley, T.
1974 *Landscape Archaeology: An Introduction to Fieldwork Techniques on Post-Roman Landscapes.* Newton Abbot: David & Charles.

Bailey, G. N., and Davidson, I.
1983 Site Exploitation Territories and Topography: Two Case Studies from Palaeolithic Spain. *Journal of Archaeological Science* 10: 87-115.

Banning, E. B.
1986 Peasants, Pastoralists and Pax Romana: Mutualism in the Southern Highlands of Jordan. *Bulletin of the American Schools of Oriental Research,* 261: 25-50.

Barker, G. and Jones, B.
1985 Investigating Ancient Agriculture on the Saharan Fringe: the UNESCO Libyan Valleys Survey. Pp 225-241 in *Archaeological Field Survey in Britain and Abroad* , eds. S. Macready and F. H. Thompson. London: Society of Antiquaries.

Barker, G. and Lloyd, J., eds.
1991 *Roman Landscapes: Archaeological Survey in the Mediterranean Region.* London: British School of Archaeology at Rome.

Bell, M., and Boardman, J., eds.
1992 *Past and Present Soil Erosion: Archaeological and Geographical Perspectives*. Oxford: Oxbow.

Binford, L.R.
1964 A Consideration of Archaeological Research Design. *American Antiquity* 29: 425-441.

Bintliff, J., and Snodgrass, A.M.
1988 Off-site Pottery Distributions: A Regional and Inter-regional Perspective. *Current Anthropology* 29: 506-513.

Borzatti von Löwenstern, E.
1994 I beduini e storia di una morfologia culturale. *Studi per l'Ecologia del Quaternario* 16: 75-89.

Boserup, E.
1965 *The Conditions of Agricultural Growth: The Economics of Agrarian Change Under Population Pressure*. Chicago: Aldine.

Bottema, S., Entjes-Nieborg, G., and Van Ziest, W., eds.
1990 *Man's Role in the Shaping of the Eastern Mediterranean Landscape*. Rotterdam: Balkema.

Bradford, J.
1957 *Ancient Landscapes: Studies in Field Archaeology*. London: Bell.

Burrough, P. A.
1994 *Principles of Geographical Information Systems for Land Resources Assessment*. Oxford: Oxford University Press.

Butzer, K.W.
1982 *Archaeology as Human Ecology*. New York: Cambridge University.

Cain, H.R.
1963 *Human and Economic Geography*. 3rd ed. London: Longman.

Cherry, J.F.
1983 Frogs Around the Pond: Perspectives on Current Archaeological Survey Projects in the Mediterranean Region. Pp. 375-416 in *Archaeological Survey in the Mediterranean Area*, eds. D. R. Keller and D. W. Rupp. BAR International Series S155. Oxford: BAR.

Cherry, J.F., Davis, J.L., and Mantzournai, E.
1991 *Landscape Archaeology as Long-term History: Northern Keos in the Cycladic Islands.* Los Angeles: University of California.

Chisholm, M.
1968 *Rural Settlement and Land Use: An Essay in Location.* 2nd ed. London: Hutchinson.

Cleave, R.
1993 *The Holy Land: A Unique Perspective.* Photography and Satellite Cartography. Oxford: Lion.

Conder, C. R.
1889 *The Survey of Eastern Palestine.* London: Palestine Exploration Fund.

Conder, C. R., and Kitchener, H. H.
1880 *Map of Western Palestine in 26 Sheets from Surveys Conducted for the Committee of the Palestine Exploration Fund, by Lieuts. C.R.Conder and H.H.Kitchener, RE, during the Years 1872-1877.* London: Palestine Exploration Fund.
1882 *The Survey of Western Palestine: Samaria.* London: Palestine Exploration Fund.
1883 *The Survey of Western Palestine: Judaea.* London: Palestine Exploration Fund.

Crawford, O. G. S.
1953 *Archaeology in the Field.* New York: Praeger.

Danby, H.
1983 *The Mishnah. Translated from the Hebrew with Introduction and Brief Explanatory Notes.* Oxford: Oxford University Press.

Dauphin, C.
1991-92 Man Makes His Landscape. *Bulletin of the Anglo-Israel Archaeological Society* 11: 22-28.

Dauphin, C., and Gibson, S.
1992-93 Ancient Settlements in their Landscapes: the Results of Ten Years of Survey on the Golan Heights (1978-1988). *Bulletin of the Anglo-Israel Archaeological Society* 12: 7-31.

Dennell, R.W.
1987 Geography and Prehistoric Subsistence. Pp. 56-76 in *Landscape and Culture: Geographical and Archaeological Perspectives*, ed. M. Wagstaff. Oxford: Blackwell.

Dever, W.G.
1988 Impact of the 'New Archaeology'. Pp. 337-352 in *Benchmarks in Time and Culture: Introduction to Palestinian Archaeology*, eds. J. F. Drinkard, G. L. Mattingly and J. M. Miller. Atlanta: Scholars.

Fiches, J-L, and Van Der Leeuw, S.E., eds.
1990 *Archeologie et Espaces: Xe Rencontres Internationales d'Archeologie et d'Histoire, Antibes, Octobre 1989.* Juan-Les-Pins: APDCA.

Finkelstein, I.,
1988 *The Archaeology of the Israelite Settlement.* Jerusalem: Israel Exploration Society.

Finkelstein, I., and Gophna, R.
1993 Settlement, Demographic, and Economic Patterns in the Highlands of Palestine in the Chalcolithic and Early Bronze Periods and the Beginning of Urbanism. *Bulletin of the American Schools of Oriental Research* 289: 1-22.

Foley, R.
1977 Space and Energy: A Method For Evaluating Habitat Value and Utilization in Relation to Archaeological Sites. Pp. 163-188 in *Spatial Archaeology,* ed. D. L. Clarke. New York: Academic Press.

Gal, Z.
1993 Regional Survey Projects: Revealing the Settlement Map of Ancient Israel. Pp. 453-458 in *Biblical Archaeology Today, 1990. Proceedings of the Second International Congress on Biblical Archaeology,* eds. A. Biran and J. Aviram. Jerusalem: Israel Exploration Society.

Gavish, D.
1990 Aerial Perspective of Past Landscapes. Pp. 308-319 in *The Land That Became Israel: Studies in Historical Geography,* ed. R. Kark. New Haven: Yale University Press.
1991 *Land and Maps: The Survey of Palestine, 1920-1948.* Jerusalem: Yad Ben Zvi (Hebrew)

Gibson, S.
1999 British Archaeological Institutions in Mandatory Palestine, 1917-1948. *Palestine Exploration Quarterly* 131: 115-143.
2001 Agricultural Terrace and Settlement Expansion in the Highlands of Early Iron Age Palestine: Is There Any Correlation Between the Two? Pp. 113-146 in *Studies in the Archaeology of the Iron Age in Israel and Jordan,* ed. A. Mazar. Sheffield: Sheffield Academic Press.

Gibson, S. and Dauphin, C.
1990 Landscape Archaeology at er-Ramthaniyye in the Golan Heights. Pp. 35-45 in *Archeologie et Espaces: Xe Rencontres Internationales d'Archeologie et d'Histoire, Antibes, Octobre 1989*, eds. J-L. Fiches, and S. E. Van Der Leeuw. Juan-Les-Pins: APDCA.

Gibson, S., and Edelstein, G.
1985 Investigating Jerusalem's Rural Landscape. *Levant* 17: 139-155.

Gibson, S., Ibbs, B., and Kloner, A.
1991 The Sataf Project of Landscape Archaeology in the Judaean Hills: A Preliminary Report on Four

Seasons of Survey and Excavation (1987-89). *Levant* 23: 29-54.

Gibson, S., Kingsley, S., and Clarke, J.
1999 Town and Country in the Southern Carmel: Report on the Landscape Archaeology Project at Dor (LAPD). *Levant* 31: 71-121.

Gibson, S., and Lass, E.
2000 Modi'in Excavation Project. *Excavations and Surveys in Israel* 20: 68*-90*

Haiman, M.
1986 *Map of Har Hamran - Southwest (198).* Jerusalem: Israel Antiquities Authority.

Harper, R. P., and Pringle, D.
2000 *Belmont Castle: The Excavation of a Crusader Stronghold in the Kingdom of Jerusalem.* British Academy Monographs No.10. Oxford: Oxford University Press.

Higgs, E. S., and Vita-Finzi, C.
1972 Prehistoric Economies; A Territorial Approach. Pp. 27-36 in *Papers in Economic Prehistory*, ed. E. S. Higgs. Cambridge: Cambridge University Press.
1975 Appendix A. Site Catchment Analysis: A Concise Guide to Field Methods. Pp. 223-224 in *Palaeoeconomy*, ed. E. S. Higgs. Cambridge: Cambridge University Press.

Isaac, B.
1990 *The Limits of Empire: The Roman Army in the East.* Oxford: Oxford University Press.

Kark, R.
1990 *The Land That Became Israel: Studies in Historical Geography.* New Haven: Yale University Press.

Kasher, A., et al., eds.
1986 *Man and Land in Eretz-Israel in Antiquity.* Jerusalem: Yad Ben Zvi (Hebrew).

Kedar, B. Z.
1999. *The Changing Land Between the Jordan and the Sea. Aerial Photographs from 1917 to the Present.* Jerusalem: Yad Ben Zvi.

Kedar, B. Z., and Danin, A., eds.
2000 *Remote Sensing: The Use of Aerial Photographs and Satellite Images in Israel Studies.* Jerusalem: Yad Ben Zvi (Hebrew).

Keller, D.R. and Rupp, D.W., eds.
1983 *Archaeological Survey in the Mediterranean Area.* BAR International Series 155. Oxford: BAR.

Kelso, W. M. and Most, R., eds.
1990 *Earth Patterns: Essays in Landscape Archaeology.* Charlottesville: University Press of Virginia.

Kennedy, D. L., ed.
1989 *Into the Sun: Essays in Air Photography in Archaeology in Honour of Derrick Riley.* Sheffield: University of Sheffield.

Kennedy, D.L., and Riley, D.N.
1990 *Rome's Desert Frontier From the Air.* London: Batsford.

Kochavi, M.
1989 The Land of Geshur Project: Regional Archaeology of the Southern Golan. *Israel Exploration Journal* 39: 1-17.

Lloyd, J. A., Owens, E. J. and Roy, J.
1985 The Megalopolis Survey in Arcadia: Problems of Strategy and Tactics. Pp. 217-224 in *Archaeological Field Survey in Britain and Abroad*, eds. S. Macready and F. H. Thompson. London: Society of Antiquaries.

MacDonald, B.
1988 *The Wadi el Hasa Archaeological Survey 1979-1983. West Central Jordan.* Waterloo: Wilfrid Laurier University.

Macready, S. and Thompson, F.H., eds.
1985 *Archaeological Field Survey in Britain and Abroad.* London: Society of Antiquaries.

Miller, N. F., and Gleason, K. L.,
1994 *The Archaeology of Garden and Field.* Philadelphia: University of Pennsylvania Press

Portugali, Y.
1982 A Field Methodology For Regional Archaeology. *Tel Aviv* 9: 170-188.
1983 Methodological Problems in Undertaking Surveys and Analyzing the Results. In *Proceedings of the Tenth Archaeological Conference in Israel*: 4. Jerusalem: Israel Antiquities Authority (Hebrew).
1987 The Settlement and its Environment: Land Use. In *Tell Qiri: A Village in the Jezreel Valley: Report of the Archaeological Excavations 1975-1977* (Qedem 24), in eds. A. Ben-Tor and Y. Portugali. Jerusalem: Israel Exploration Society.
1993 Regional Excavation Projects. P. 452 in *Biblical Archaeology Today, 1990. Proceedings of the Second International Congress on Biblical Archaeology,* eds. A. Biran and J. Aviram. Jerusalem: Israel Exploration Society.

Riley, D.N.
1980 *Early Landscape From the Air.* Sheffield: University of Sheffield.

Roberts, B.K.
1987 Landscape Archaeology. Pp. 77-95 in *Landscape and Culture: Geographical and Archaeological Perspectives,* ed. M. Wagstaff. Oxford: Blackwell.

Roper, D.C.
1979 The Method and Theory of Site Catchment Analysis: A Review. Pp. 119-140 in *Advances in Archaeological Method and Theory.* Vol.2, ed. M. G. Schiffer. New York: Academic Press.

Rosen, A.
1986 Environmental Change and Settlement at Tel Lachish, Israel. *Bulletin of the American Schools of Oriental Research* 263: 55-60.

Rosen, S.
1987 Byzantine Nomadism in the Negev: Results from the Emergency Survey. *Journal of Field Archaeology* 14: 29-42.
1992 Nomads in Archaeology: A Response to Finkelstein and Perevolotsky. *Bulletin of the American Schools of Oriental Research* 287: 75-85.

Rossignol, J., and Wandsniper, L.,
1992 *Space, Time, and Archaeological Landscapes*. New York and London: Plenum Press.

Schiffer, M.G., ed.
1979 *Advances in Archaeological Method and Theory*. Vol. 2. New York: Academic Press.

Schofield, A.J., ed.
1991 *Interpreting Artefact Scatters: Contributions to Ploughzone Archaeology*. Oxford: Oxbow.

Tiffany, J.A., and Abbot, L.R.
1982 Site-catchment Analysis: Applications to Iowa Archaeology. *Journal of Field Archaeology* 9: 313-322.

Ucko, P. J., and Layton, R.
1999 *The Archaeology and Anthropology of Landscape: Shaping Your Landscape*. London and New York: Routledge.

Vallat, J-P.
1991 Survey Archaeology and Rural History: A Difficult but Productive Relationship. Pp. 10-17 in *Roman Landscapes: Archaeological Survey in the Mediterranean Region*, eds. G. Barker and J. Lloyd. London: British School of Archaeology at Rome.

Van Andel, T. H., and Runnels, C.
1987 *Beyond the Acropolis: A Rural Greek Past*. Stanford: Stanford University Press.

Van Andel, T. H., Runnels, C.N., and Pope, K.O.
1986 Five Thousand Years of Land Use and Abuse in the Southern Argolid, Greece. *Hesperia* 55: 103-128.

Vita-Finzi, C.
1978 *Archaeological Sites in Their Setting*. London: Thames and Hudson.

Vita-Finzi, C., and Higgs, E. S.
1970 Prehistoric Economy in the Mount Carmel Area of Palestine: Site Catchment Analysis. *Papers of the Prehistoric Society* 36: 1-37.

Wagstaff, M., ed.
1987 *Landscape and Culture: Geographical and Archaeological Perspectives*. Oxford: Blackwell.

Wagstaff, M.
1988 The Tenta Catchment: Resource Exploitation in Neolithic Cyprus. Pp. 64-71 in *Conceptual Models in Environmental Archaeology*, eds. J. Bintliff, et al. Edinburgh: Edinburgh University Press.
1991 The Archaeological Site From a Geographical Perspective. Pp. 9-10 in *Interpreting Artefact Scatters: Contributions to Ploughzone Archaeology*, ed. A. J. Schofield. Oxford: Oxbow.

Webley, D.
1972 Soils and Site Location in Prehistoric Palestine. In *Papers in Economic Prehistory*, ed. E. S. Higgs. Cambridge: Cambridge University Press.

Wilkinson, T.J.
1991 Landscape Archaeology in the Near East. *BANEA Newsletter* 4: 17-18.

Wilson, D.R.
1982 *Air Photo Interpretation For Archaeologists*. London: Batsford.

Zertal, A.
1992 *The Mansseh Hill Country Survey: The Shechem Syncline*. Haifa: The University of Haifa Press (Hebrew).

Zertal, A. and Greenberg, M.
1983 Methodological Problems and Simulation Studies in Archaeological Survey. Pp. 3 in *Proceedings of the Tenth Archaeological Conference in Israel*. Jerusalem: Israel Antiquities Authority (Hebrew).

Figure 1-1: Terraced landscape in Wadi Battir, south-west of Jerusalem. Photograph by Krikorian taken in 1892.

Figure 1-2: Royal Air Force plane over Bethlehem in a photograph taken in 1936.

Figure 1-3: Aerial photograph of Mazra'at Kuneitra in the Golan Heights. Note the well organized walls of a co-axial field system dating from the Roman period. Photograph taken in the 1970s.

Figure 1-4: The Spring of el-Khaniyyeh in the Rephaim Valley. Photograph by Henry Phillips, RE, taken in 1867.

Figure 1-5: Ancient road (probably Roman) near the former village of Imwas (Emmaus). Photograph by Derrik Riley taken in 1990.

Figure 1-6: One of the Roman military camps at Battir with later terraces and houses built over it. Photograph by Derrik Riley taken in 1991.

Figure 1-7: General view of agricultural terraces at Sataf.

Figure 1-8: Area B of the excavations at Sataf. The EB terrace is in the center of the excavated area.

Figure 1-9: The large plastered collecting pool at Ain Sataf.

Figure 1-10: Farj: one of the stone boundary walls of the Roman field system.

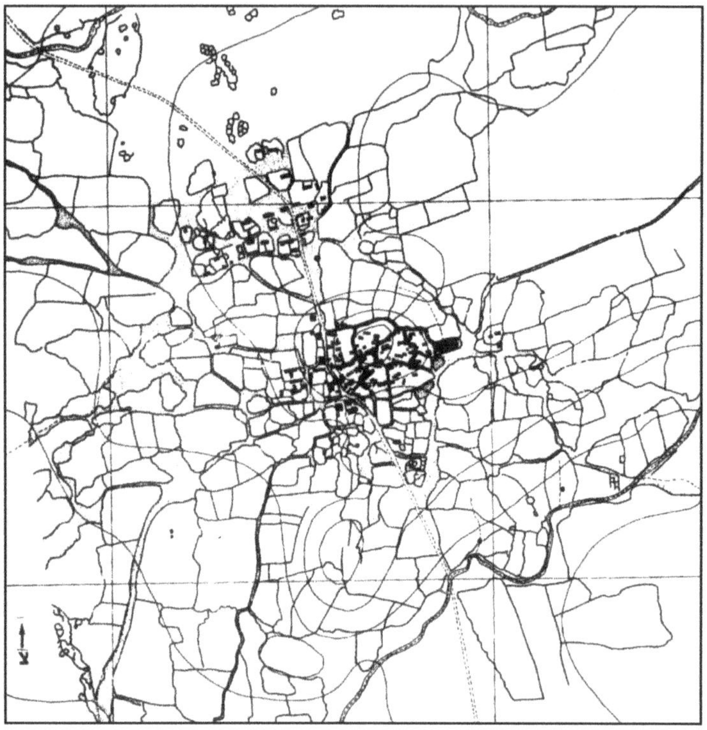

Figure 1-11: Ramthaniyye: map of the settlement and its surrounding fields.

Figure 1-12: Ramthaniyyeh: one of the enclosures from the Byzantine period.

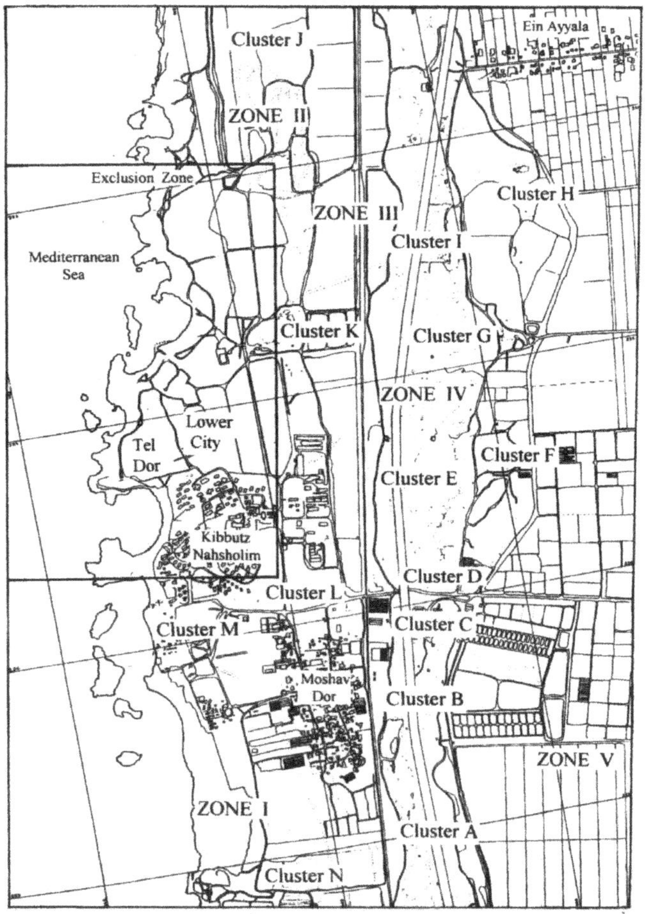

Figure 1-13: Dor: map of the surveyed area.

Figure 1-14: Modi'in: curvilinear rural structures dating from the Early Roman period.

Figure 1-15: Modi'in: oval stone clearance heap.

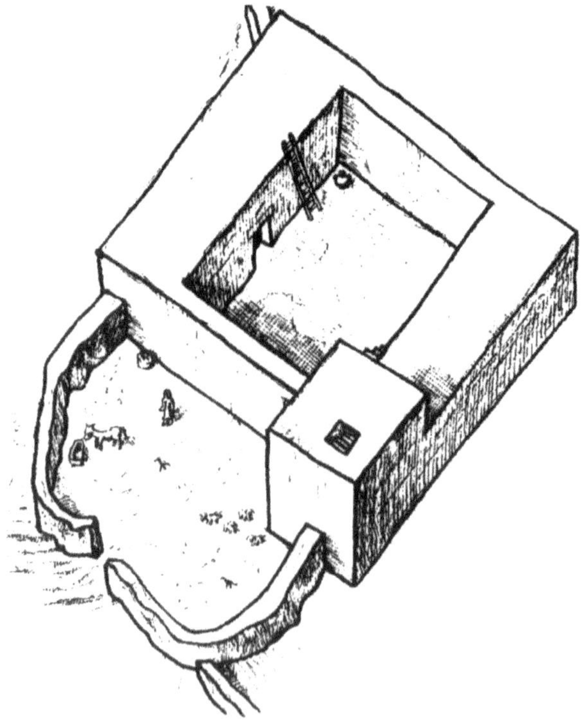

Figure 1-16: Modi'in: reconstruction of the farm building dating from the late 2nd century BC.

Figure 1-17: Modi'in: the interior of an ancient charcoal burner in the process of being recorded.

Discontinuities in Rural Settlement in Early Bronze Age – Middle Bronze Age I Palestine

RAPHAEL GREENBERG

Department of Archaeology and Ancient
Near Eastern Studies
Tel Aviv University
P.O. Box 39040
Tel Aviv, 69978
Israel
grafi@post.tau.ac.il

In several recent studies, rural life in Early Bronze Age Canaan has been depicted as a permanent component of the settlement landscape. This fixed element, to which the Palestinian landscape is "preadapted," was, in these views, affected by occasional evolutionary and devolutionary cycles, leading to and away from urbanism. Following the lead of Portugali and Gophna, and using data from the Hula valley in northern Israel, the presumed stability of rural settlement in the Early Bronze Age will be called into question. Do EB IA hamlets indeed develop into EB IB villages and these into EB IB/II towns? Is a fundamentally stable rural network maintained throughout the urbanizing phases of EB II and EB III? Is this the same network that re-emerges at the end of EB III to form the basis for Intermediate Bronze Age and MB I culture? An alternative view will be proposed, according to which rural life was subject to significant discontinuities. Some hypotheses explaining these discontinuities will be offered, at least as points for future discussion.

In an article published in 1979, Leon Marfoe provided a reasoned formulation of what was to become the primary model for the description of early social formations in Palestine: the pastoral–urban continuum. The physically fragmented environment of the southern Levant, Marfoe argued, was particularly conducive to small-scale social organization and to a politics of fragmentation. For the most part, communities maintained a flexibility that allowed them to oscillate between different modes of subsistence without dissolving basic ties of kinship and general social bonds. Larger organizations came into being, and were precariously maintained, only when environmental conditions permitted the creation of stable economic relationships and when elites were able to enforce their centralizing tendencies upon local groups. Their success was dependent on the degree to which they were able to provide ideological motivation for centralization. This ideology was perceived by Marfoe as going against the grain of the centrifugal tendencies of a basic subsistence economy, and of the small-scale social units that practiced it (Marfoe 1979).

The "pastoral-urban continuum" quickly took center stage in the analysis of social transitions in Palestine, particularly in the study of the Early Bronze–Middle Bronze transition. As the older formulations invoking large-scale migration and military invasion increasingly fell out of favor, the depiction of indigenous society as flexible, resilient, and able to change its external form while maintaining its internal essence grew increasingly attractive (e.g., Richard 1987; Finkelstein 1991; Dever 1995; Harrison 1997). An elastic ruralism came to be viewed as the most successful adaptive response to external conditions, not only in the late third millennium, but throughout the Early Bronze Age. This approach was increasingly championed by W.G. Dever and his students at the University of Arizona (henceforth, the Arizona school). The most explicit presentation of this view of the village as the fundamental building block of society was put forth by Joffe (1991, 1993). Joffe viewed the southern Levantine countryside as "preadapted" to village settlement, which underwent recurring cycles of rising complexity (evolution) and systemic collapse. In an extensive study of the Intermediate Bronze Age, G. Palumbo (1991) went so far as to state that "the typical rural settlement of the EB I, autonomous and basically independent from the influence of any kind of 'central places', continued its existence through periods of different levels of specialization ..." into EB IV (Palumbo 1991:128), and even beyond (Palumbo 1991:62). Implicit in this essentialist view of rural life was the relegation of urban organization to the status of an evolutionary non-starter. In this vein, Dever stated baldly that "the phenomenon to be explained is not the collapse of the Palestinian urban EB II-III culture, but rather its *existence* in the first place" (Dever 1989:238); Palumbo (1991:6) characterized Early Bronze urbanism as an "ultimately unsuccessful 'experiment'," a sentiment echoed by Falconer (1994:308).

This attitude toward urbanism was not maintained without diffidence: The apparent discrepancy between the impressive record of third millennium urban life and its theoretical disparagement could not be ignored. One solution was to redefine urbanism so that few or none of the Early Bronze Age walled settlements would qualify as towns (Falconer 1994:312), thus reducing the scale of social change to oscillation between large and small villages. A more sophisticated approach (Joffe 1993:90–93) returned, ultimately, to the infrastructure-superstructure dichotomy suggested by Marfoe: Internal contradiction between power-wielding elites and resisting villagers and nomads is a permanent characteristic of urban settlement in Canaan. In the long run, the rural forms of existence are the more durable (Joffe 1993:85).

A radical alternative to the Arizona school was offered by Portugali and Gophna (1993), based on earlier theoretical work by Portugali (1989; 1994). They maintained that it was rural society which was prone to crisis and constantly on the brink of dissolution. Urban configurations, in which towns play the crucial role of integrating the economy, were able to absorb short-term environmental fluctuations and maintain long-term stability. When these 'global' configurations did collapse, however, the ensuing crisis was of far greater intensity than that characterizing rural societies (Portugali and Gophna 1993:166–167; Portugali 1994). By implication it may be understood that the nature of rural life itself was altered by the significant changes in the relations of production that accompanied urbanization. This alternative view has yet to receive the attention it merits, perhaps because it was appended to a speculative model of urbanization inadequately grounded in the archaeological evidence (Finkelstein 1995; but see Finkelstein and Ussishkin 2000:585).

The two approaches outlined above represent divergent views on the causes and mechanisms of change in society. The former approach views settlement history as a series of cycles of evolution and devolution. Societies are envisioned as being on a trajectory, leading to or away from increased (and fundamentally unstable) complexity. Thus, Joffe views the EB IA villages teleologically, as a first, necessary step on the path leading to the creation of urban life, "the launching point of the larger evolutionary trajectory" (Joffe 1993:48). In contrast, Portugali appears to view settlement history as a disjunctive process, in which changes occur as a result of "external or internal stimuli which force the system to independently self-organize" (Portugali and Gophna 1993:182).

Although many of the archaeological details brought in support of their reasoning are problematic, Portugali and Gophna's approach rectifies two basic flaws in that of Marfoe and the Arizona school: a conception of village life as essentially static, and an artificial division between village infrastructure and urban superstructure. The view that accords rural existence an essential adaptive superiority raises the question of why the urban option should have been maintained for such extensive periods of time, apparently against the better judgment of the population at large. Furthermore, it does not allow for the possibility that the new relations of production accompanying urbanization entailed a fundamental restructuring of life in villages as well as in towns.

Although these divergent views seem to draw on the same, limited body of evidence, they are not necessarily equally valid, and they should be susceptible, to some extent, to archaeological confirmation. Closely dated settlement maps and stratigraphic sequences can reveal patterns of continuity or discontinuity. The degree of urban integration in the countryside, and of interaction between town and village, can be estimated. Crises on a global scale should be evident in the record. Contacts with foreign cultures at crucial transitions ought to be identified. Most importantly, long-scale trends and trajectories should have some verifiable, short-term, archaeological correlates. An impressionistic canvas that dissolves into mere blotches of color upon close inspection is an inadequate metaphor for archaeological inquiry. Rather, we should seek to attain both the distant perspective and accurate detail of an aerial photo.

The various studies cited above fall short of the mark. All are based on fragmentary evidence culled from various parts of Palestine and blended together to create a uniform model. Surveys are cited that almost invariably employ an extremely coarse-grained chronology, dividing the 12-odd centuries of the Early Bronze Age into two unequal parts: EB I on the one hand, and EB II-III on the other. Little consideration is given to possible differences between the progress of urbanization and deurbanization in sub-regions of Palestine. In short, the models proposed by Richard, Palumbo, Joffe, and others on the one hand, and Gophna and Portugali on the other, need to be put to a more rigorous test.

The Hula Valley

The 200 km^2 Hula valley was densely settled over considerable stretches of history. Two of the most important mounds of ancient Palestine—Hazor and Laish/Dan—are located in the valley, and several important routes connecting Syria and Palestine run through or near it. A fertile, well-watered basin, surrounded by relatively difficult terrain, the Hula constitutes an ideal subject for close archaeological scrutiny: it is a region with clear geographic borders, small enough for intensive study, yet varied enough in terrain and rich enough in settlement to allow for observation of long-term change. The perennial swamps north of the now-drained Lake Hula, and the tendency of low-lying areas on the valley floor to flood during the rainy season, appear to have been the main cause of extreme oscillations in the density of settlement in the valley. While the environment was conducive to settlement and could support a dense population, its degradation by human agents could rapidly transform it into a marginal area. This makes the valley a sensitive barometer of settlement trends in Palestine as a whole.

The chronostratigraphic and cultural sequence of the Hula valley in the mid-fourth to early second millennium has been established with the aid of a number of published excavations and fairly comprehensive surveys.[1] Early Bronze I has been studied in detail at Tel Te'o (Eisenberg 1989; Eisenberg, Gopher and Greenberg, 2001) and at the Gadot tomb (Greenberg, 2001). Chance finds and surveys have led to the discovery of numerous additional sites (Greenberg 1990; 1996b; Stepansky 1992; 1994; 1999; Greenberg et al. 1998; Shaked and Marder 1998). These surveys have also revealed a picture of dense EB II settlement, and are augmented by the publication of stratified EB II–III sites (Tel Dan, in Greenberg 1996a; Tel Hazor in Greenberg 1997). Intermediate Bronze Age occupation has been identified in tombs (Amiran 1961; Yeivin 1966; Eisenberg 1985; Covello-Paran 1996), mounds (Yadin 1972:120–121; Biran 1994:45) and scattered settlement sites. Early second millennium occupation has been studied at Tel Dan (Ilan 1992; 1996; Biran 1994:47–90), Tel Hazor (Maeir 1997), Tel Na'ama (Greenberg et al. 1998), Kefar Szold (Epstein 1974), Hagosherim (Covello-Paran 1996) and in the surveys mentioned above.

These studies provide a detailed cultural sequence for the period in question, permitting the translation of excavation and survey results into discrete settlement maps for EB IA, EB IB, EB II, EB III, IBA, and MB I. Of course, the maps are reifications, a static picture of what were dynamic episodes of settlement, each extending over 150–400 years. Thus appearances can be misleading, and the sites placed on a given map may not have been truly coeval. Excavated stratigraphic sequences within the periods could provide increased resolution for certain segments of space and time, but only further research will help create a truly fine-grained portrait.

Early Bronze Age I

Nineteen settlement sites and one tomb are attributed to the EB I (Fig. 2-1). Many of these sites were identified by chance, in the course of development projects, because they typically are located on the western valley floor, in areas affected by later alluviation. It may thus be assumed that they are under-represented in the archaeological record. Nearly all the sites have ceramic evidence of early to mid-EB I occupation (EB IA). Only one site was definitely occupied in EB IB.

Tel Te'o, which has been excavated (Eisenberg 1989), is probably, at 3–4 ha, the largest of the EB I sites in the valley. A tomb excavated at Gadot (Greenberg 2001) and the intensively collected site of Kh. Khisas (Greenberg 1996b:63–5) provide significant pottery assemblages, complementing that of Tel Te'o. Gray Burnished ware of the early type (Wright 1958, Type 1) has been found at nine of the sites, and red-slipped pottery in all of them. Most typical of the Hula valley sites are the sloped-shouldered jars with inserted upright or flared rims. Conspicuous in their absence both in the Gadot tomb and at the settlement sites are typical EB IB forms: late Gray Burnished ware, modelled-rim bowls, ridged holemouth jars, ledge-handled jars with splayed rim, bent-spout teapots, and other components of what Schaub (1982) has termed the "B tradition." At three sites on the eastern flank of the valley several "grain-wash" sherds were found. Long thought to be diagnostic for EB IB, recent studies have indicated that grain-wash begins in the middle of EB I and continues into EB III in and around the Jordan valley (Genz 2000; Kamlah 2000). Diagnostic EB IB pottery—fragments of rail-rim pithoi—has been found at only one site (Nahal Quba'at), at the southwestern tip of the valley.

Tel Te'o, the largest site in the valley, appears to have been comprised of loosely-aligned oval houses (Fig. 2-2)—a plan well attested in EB IA in different parts of the Levant. There is no evidence that it served as a "central place" vis-à-vis the other sites, none of which seem to have exceeded 1 ha. in size. In fact, its location within a narrow strip bordered by the Hula swamp on the one side and the steep slopes of the Galilee hills on the other was hardly advantageous for such a function. Notably, neither of the two EB II–III urban sites—Tel Dan and Tel Hazor—was occupied in EB I.

The location of many of the EB I sites on the floor of the valley—in contrast to later periods—suggests that the value of agricultural land was not at a premium at this time. Similar sites of this period, such as Yiftah'el (Braun 1997) or Sidon-Dakerman (Saidah 1979), appear to show horizontal stratigraphy, i.e., the gradual lateral displacement of domestic structures over time. Such choices of location and spatial organization are characteristic of societies in which individual or hereditary property rights are not yet established. Communal exploitation of land within the framework of a mixed Mediterranean subsistence economy, therefore, seems a reasonable characterization of the socio-economic structure of the Hula Valley villages (cf. Joffe 1993:48).

Early Bronze Age II

Eleven of the eighteen EB IA sites abandoned in EB IB, as well as the one EB IB site, were occupied in EB II (Fig. 2-3). The Gadot tomb was also reused after an interval represented by the accumulation of sediment in the tomb. Eleven new sites were established in this period. In addition to the temporal discontinuity and partial abandonment of site locations, there is a major shift in the center of gravity of settlement: most sites are located along the eastern flank of the valley, and the most important EB II sites —Tel Dan, Tel Hazor, Tel Abel and Tel Khisas—were established on virgin or long-abandoned ground (e.g., Tel Dan had last been settled during the Pottery Neolithic). The major EB I site at Tel Te'o shows only minor rehabitation during EB II.

The pottery from all EB II sites is characterized by a significant component of North Canaanite Metallic Ware (Greenberg and Porat 1996). Excavated sites include Tel Dan (three stratigraphic phases), Hazor (two or three phases), Tel Te'o (one phase), and the Gadot tomb. Test pits at sites such as Shamir and Tel 'Ateret suggest single-phase EB II occupation.

The sites utilized a wide variety of ecological niches: alluvial terraces, topographical prominences on the plain, hill-slope terraces, and, especially, transitional points to other regions (Dan, Abel, Hazor, and 'Ateret). The alluvium itself, however, was generally left free for cultivation. There is significant variation in the size of sites, suggesting a three-tiered hierarchy, with the three main gateway sites of Dan, Abel and Hazor at 10–20 ha., large villages at 1–4 ha., and small villages at under 0.5 ha. An early attempt to arrange the sites on a central-place grid (Greenberg 1990) proved unsuccessful in predicting the location of subsequently identified sites. For example, the fertile 30 km^2 plain between Hazor and Lake Hula (where three sites were located in EB I) is unoccupied in EB II, whereas sites along the eastern and northwestern slopes of the valley are located within 1–3 km of each other, in an unexpectedly dense linear pattern. A more promising avenue for locational analysis appears to be the arrangement of sites along major routes transecting the valley and leading to neighboring regions of contemporaneous settlement (in the Golan, Galilee, Beqa', and central Jordan Valley). This may be characterized as a dendritic pattern of settlement, typifying colonization processes (Hirth 1978).

The stratigraphy of the large sites and the uniform material culture of all the sites can be interpreted to indicate that Dan and Hazor were settled first, and served as nodal points for the swift colonization of the entire valley. A short time after the reestablishment of settlement, a web of hierarchically arranged sites occupied the area. The 22 sites are estimated to have occupied approximately 80 ha, with a population possibly numbering 10,000 souls. Such a population would have required not only intensive cultivation of the valley floor for subsistence, but maintenance of close economic ties with inhabitants of the margins of the valley, for supply of meat and dairy products (cf. Rosen 1994). Continuous effort would also have been required to maintain agricultural productivity: ground clearance and delineation of property rights (and perhaps irrigation rights as well) on the now valuable land, and the provision of adequate drainage to prevent the encroachment of marshes.

The density of settlement and the need for economic integration and control surely implies a concentration of authority. The most tangible expression of such a centralized economic authority is the production and dissemination of North Canaanite Metallic Ware. This ware, probably manufactured on the southern slopes of Mt. Hermon (Greenberg and Porat 1996), is the dominant ceramic artifact in the central Levant in EB II, as far west as the Mediterranean coast, as far south as the southern Jezreel valley, and as far east as the Hauran. Its uniform production and wide distribution reflects not only extreme specialization and centralization, but also an ideological unity prevalent in large parts of the countryside, emanating from the Hula Valley towns (Greenberg 1999). Local communities forwent the privilege of producing their own ceramics, finding their cultural codes sufficiently represented in Metallic Ware. The discovery of quantities of North Canaanite Metallic Ware in royal First Dynasty contexts in Egypt (Kantor 1992) indicates the important part played by international trade in the rise of the urban polity of North Canaan (Greenberg and Eisenberg, forthcoming).

Hula Valley EB II settlement may thus be characterized as an agricultural-commercial complex, with integrative connections between its constituent settlements. The development of this complex has none of the characteristics of gradual urban evolution. Urbanism appears to have arrived ready-made in the valley from an adjacent region, probably the central Jordan valley, where urbanization seems to have come early. The Hula Valley villages, therefore, were not expressions of a permanent, long-lasting rural component, but intentional creations of a colonizing movement, and were highly dependent on the towns for their existence.

Early Bronze Age III

This period is marked by a sharp drop in the number of settlements: only three (Dan, Hazor and Khisas) are identified with certainty (Fig. 2-4), while Abel may also have been occupied (Dever 1986). If these four large sites indeed coexisted, they would represent approximately 70% of the areal extent of EB II settlement, indicating a shift in settlement toward agglomeration, rather than mere abandonment. Tel Dan and Hazor were major towns in EB III, evincing considerable interaction with other parts of northern Canaan, and thus constituting part of a larger web of cities of this period (Finkelstein 1995). Clearly, however, the rural hinterland of these two towns was severely diminished.

The florescence of the EB II agricultural-commercial complex was thus evidently short-lived. Gone too was the cultural integration of EB II. During EB III—as indicated by the material culture of Dan and Hazor (Greenberg 2000)—each city developed along an independent trajectory (see, e.g., Greenberg 1997:191). The introduction of fine-wares, including—but not limited to—Khirbet Kerak ware, suggests a greater degree of social differentiation within the cities. Metallic ware became only one of a series of locally-produced pottery styles, and the Hula valley lost its preeminence in inter-regional trade in agricultural goods. Huge fortifications were constructed at Tel Dan (Biran 1994:68–70), reflecting, on the one hand, the organizational power of a military-political elite, but conveying, on the other, a sense of increasing insecurity and isolation.

The following scenario may explain the disintegration of the EB II complex, the emergence of EB III city-states, and their eventual collapse:

1. As suggested above, the existence of the EB II settlement complex depended on the maintenance of both ecological and social balances. The former included maintaining agricultural productivity by continuous investment in drainage and land clearance; were these to be neglected, a swiftly spiralling decline in productivity would ensue. The latter included the promotion of solidarity between the inhabitants of both city and countryside, and the curbing of elite self-aggrandizement.
2. Initially, as long as the system expanded, these balances were maintained. Once settlement exploited every available niche in the valley, however, divergent interests began to come into play. Economic elites found in the rural hinterland a potential for enlarging institutional (and possibly individual) holdings, at the expense of small landowners. A process well-documented in other regions and eras was set into motion, leading eventually (a) to the formation of large city-owned holdings and a class of dispossessed sharecroppers, some of whom would have found their way to the semi-nomadic periphery, and (b) to competition and the acceleration of centrifugal tendencies among the large towns (cf., e.g., Adams and Nissen 1972:17–21; Rowton 1976; Marfoe 1979:20–23; Millon 1988).
3. As the rural hinterland became a staging ground for inter-urban rivalry, villages were abandoned. The accelerated growth of urban institutions, presumably centered around palace and temple (Herzog 1997: 94–97), allowed towns to grow in size and population. Their extensive fortifications advertised increased security during increasingly unsettled times, and the formation of large estates by the urban elite may initially have acted as a buffer against agricultural bad years (Postgate 1994:186).
4. With time, the abandonment of the hinterland led to the encroachment of swamps on the productive areas, the rise of disease, and the spiralling decline of productivity. The towns became increasingly vulnerable to economic crisis, and eventually reached the point at which the investment required to maintain urban institutions outweighed their perceived advantage, i.e., the brink of collapse (Tainter 1988).
5. In what may have been a domino effect linked to developments in other parts of Canaan, towns began to abandoned. Urban collapse brought about yet another dramatic shift in the settlement map of the valley—the virtual abandonment of mounds and the dispersal of settlement, as described below.

Dolmens

A belt of dolmen fields surrounds the northern Hula Valley (Figs. 2-4, 2-5), occupying areas previously settled in EB II. These fields include both basalt and limestone dolmens, in accordance with the local lithology. A number of dolmens have been excavated, but none are accurately dated (Shaked 1994, 1995, 1999). A connection could be posited between the abandonment of the two lower EB II settlement tiers in EB III and the construction of dolmens. These megalithic structures might reflect the semi-nomadization of former village populations along the margins of the valley and could represent a mode of territorial demarcation among seminomadic groups or between semi-nomads and towns. Such semi-nomadic populations may well have remained in the area—and continued to build dolmens—in the Intermediate Bronze Age.

Intermediate Bronze Age

Eleven sites of this period—seven habitation sites and four tomb sites—were identified (Figure 2-5), in addition to the dolmen fields, which were probably in use at this time (Bahat 1972; Epstein 1985). Three habitation sites and two tombs were located on the valley floor, suggesting agricultural populations. The remainder were on the valley fringes, and might indicate a more moblile type of occupation. Of the habitation sites, only Tel Na‘ama shows evidence of architecture and stratified accumulation (Greenberg et al. 1998), while Dan (Biran 1994:44–45) and Hazor (Yadin 1972:120–121; Ben-Tor 1998) appear to have been occupied by "squatters," with only meager architecture and ceramic remains in evidence. The material culture recovered from the sites and tombs is remarkably heterogeneous. Local pottery types show evidence of various traditions (Eisenberg 1985; Greenberg et al. 1998). Syrian-type black wheelmade ware is common at Tel Na‘ama, and appears in the tombs of Ma‘ayan Barukh and Hagosherim (Amiran 1961; Covello-Paran 1996), but is absent at ‘Enan (Eisenberg 1985). The last site, however, contains a unique collection of tin-bronze daggers, also suggesting close connections with Syria (Stech, Muhly, and Maddin 1985). Taking into consideration the varied geographic locations, the ephemeral quality of many of the settlements, the apparent use of dolmens, and the variety of tomb gifts, the Hula Valley population seems to have been very limited in size and to have practiced varied modes of subsistence, representing a sharp break with earlier periods.

The combination of insubstantial settlement and imported preciosities suggests that the Hula Valley was in this period a margin of the Syrian periphery, which, in turn, was closely interactive with a continuously urban Syrian core area (see, e.g., Dever 1980). The dissolution of the Hula valley towns—the end result of the spiralling decline described above—would have resulted in the exodus of urbanized elites, the ruralization of the remaining town-dwellers, and the spread of mobile populations. The last would have come into occasional contact with more sedentary groups in Lebanon or southern Syria. Such contacts were marginal, in the sense that they did not generate any form of social change within the Hula Valley during the final centuries of the third millennium BCE.

As margin to an urban core, the hamlets and encampments of the Intermediate Bronze Age cannot be seen as a reemergence of a rural substratum in the Hula valley. These types of occupation are post-urban, rather than merely non-urban or pre-urban, and are not really comparable in form or substance to any earlier settlement.

Middle Bronze Age I (traditional Middle Bronze Age IIA)

Twelve sites were attributed to this period, with five additional sites tentatively ascribed an MB I date (Fig. 2-6). Of the twelve, most were established on sites previously settled during the Intermediate Bronze Age (Tel Dan, Tel Hazor, Na‘ama, Hagosherim tombs) or earlier.

The distribution of the sites in the landscape was balanced: one quarter on the alluvial plain, and the remainder scattered on terraces and slopes, with no clear preference for one or the other side of the valley. There is no evidence of a concentration of settlements along routes, as in EB II, or a preponderance of large towns, as in EB III. Rather, there seems to be a very well-spaced distribution of sites of various sizes. More than in any other period, the MB I seems to show the gradual development of a balanced settlement system, with central places emerging over time, first at Tel Dan (Ilan 1996:163–165), and later—perhaps at the very end of the period, as settlement spread southward—at Tel Hazor (Ben-Tor 1997:6–7).

The perception of the Hula Valley as a permeable margin of northern culture-centers in the Intermediate Bronze Age is a key to understanding processes that occurred in MB I. The earliest remains of the latter period are scant, consisting of mortuary finds and ephemeral settlement remains at Tel Dan, reused Intermediate Bronze Age tombs at Hagosherim (Covello-Paran 1996) and possible architectural remains at Tel Na‘ama (Greenberg et al. 1998). Neither these remains, nor the quality of the material culture found at these sites, are indicative of any dramatic shift in lifestyle, despite some innovations in pottery and tomb types. Evidence for more sedentary settlement, and eventually for reurbanization, comes later in MB I, with cities finally being established only toward the end of the 250-year-long period (Maeir, this volume). The process of resettlement by sedentary agriculturalists seems to have been gradual, very unlike EB II, and akin to classic central place models (Blouet 1972; Hagget, Cliff, and Frey 1977:157 ff). Sites seem more evenly distributed in the landscape, and village autonomy is suggested by the varied ceramics and heterogeneous clay sources indicated by petrographic analysis (Greenberg et al. 1998: Figs. 17, 18).

Generally speaking, the Hula Valley finds accord well with historical reconstructions like that of N. Na'aman (1982): the initial stage of semi-sedentary settlement, which basically continued Intermediate Bronze Age social organization, was followed by gradual agglomeration and the eventual organic growth of cities. Cultural contacts with Syria were very prominent, and the impulse toward urbanization must also have come from the north. By the end of the period, the population had grown from several hundred to some 7000 souls, and the stage was set for the emergence of the great city-state of Hazor in MB II.

Discussion

Taking a long view of Hula Valley settlement history, it may be characterized as one of fundamental discontinuity. EB IA village settlement represents the final episode of sedentary, non-stratified subsistence-level settlement that began with the Natufian of Mallaha (Eynan) and continued with extensive, but discontinuous, PPN (e.g., Beisamun, Tel Te’o), post-Yarmukian PN (e.g., Tel Te’o, Tel Dan), and Ghassulian Chalcolithic (e.g., Tel Te’o, Tel Turmus) manifestations. Rather than the starting-point of a trajectory toward urbanism, EB IA was an evolutionary dead-end, followed by general abandonment. EB II rural settlement was established as part of a large-scale urban colonization of the valley, in which villages were economically integrated with towns—a “vertical” integration, characterized by relations of dependence, rather than a “horizontal” integration, characterized by reciprocal relations. While some EB II villages were founded on EB I sites, others were not, and the general pattern of settlement is markedly different, due to the increasing intensification of cultivation on the valley floor. Urbanism and urban ideology were not foisted on these villages, but were part of their very *raison d'etre*. In order to persist over six or seven centuries, urban ideology—including a conception of town–village relations—had to become a “lived” or embedded, unconscious ideology.[2] As such, it would have left an indelible imprint on all members of society, including those in the rural hinterland.

EB II villages were relatively short-lived, and most did not survive into EB III. The latter period may be characterized as the zenith of Early Bronze urbanism; a phase of agglomeration that resulted in the depopulation of the rural countryside. The vacuum in sedentary settlement in areas between the towns probably began to be filled at this time with non-sedentary populations—pauperized agriculturalists, social outcasts, or nomads from neighboring regions. These groups, marginalized during EB III and representing the ideological rival or ‘antisystem’ to urbanism (Eisenstadt 1988:241), probably played an important role in urban collapse, although this is beyond the purview of this study. They formed a significant facet of subsequent Intermediate Bronze Age settlement, which is characterized by small-scale sedentary settlement, dolmen-building, and cultural fragmentation. This segmentary mode of settlement was no mere reversion to an older mode village life; it has nothing in common with the rural settlement of EB II, and bears little or no resemblance to the EB IA occupation. It seems, however, to have formed the basis for a gradual intensification of sedentary settlement, leading to the eventual emergence of central places in MB I. This pattern of intensification, indeed a "trajectory toward urbanism," is in turn, with its marked Syrian component, different from all earlier patterns of settlement, though it might contain elements found in them.

Is the Hula Valley evidence sufficient to invalidate the "fundamental rural stability" school of thought? The following objections may be raised:
1) The micro-regional approach may be too narrow. The valley represents a limited deviation from the norm, characteristic of areas at the margin of permanent settlement.
2) The very fluidity of settlement in the valley is the best illustration of rural resilience: each pattern represents a new adaptation of rural life to changing conditions. Continuity does not require permanence of occupation at a given site.

These objections may be countered by returning to the issues raised in the introductory paragraphs. If a generalization is made concerning EB IA–EB IB–EB II continuity and evolution, it should not dissolve into discontinuity and disjunctive change on closer scrutiny. In fact, the changes evident in Hula valley settlement, although dramatic, are not different in kind from what is found in other parts of the country. The nineteen EB IA sites of the Hula valley can hardly be written off as marginal, in terms of the EB IA settlement in Palestine as a whole. Nor is the Hula the only region that exhibits EB IA–B discontinuity. Other important EB IA sites are abandoned at the end of the period, such as Yiftah'el, Meser, and Jebel Mutawwaq, to name but a few. As for EB IB, Portugali and Gophna have documented extensive abandonment of sites (1993: Fig. 3). Little architectural and cultural continuity is observed in the EB IB–II transition at sites such as Megiddo or Bet Shean, which were abandoned, or sites such as Ai, Tel Qashish, or Rosh Haniqra, which reveal clear stratigraphic disconformities and cultural disjunction between the earlier villages and the later walled settlements. Widespread abandonment of small sites at the end of EB II has been observed in regional studies of the Negev (Beit-Arieh 1981) and the plain of Lod (Gophna 1989). Other regional studies show similar development, though it is often obscured by insufficient chronological resolution. In short, if not a paradigm for the settlement history of Palestine as a whole, the Hula Valley evidence should at least be seen as an expression of an alternative mode of development, and an antidote to facile generalizations. As P. de Miroschedji has put it, "il n'y a pas une Palestine, mais plusieurs" (de Miroschedji 1989:73): Each region studied will show its own distinctive settlement history within the larger trend, and disjunctive change seems to characterize many of these regions.

As for rural resilience, the question here is one of significance. If what is implied is that virtually all modes of settlement in Palestine are based on a mixed Mediterranean economy adapted to local environmental constraints and taking on different social forms—there is hardly room for objection, except on grounds of triviality. What I have tried to show is that changes in social complexity significantly affect the economic organization of rural society itself, the way in which the landscape is utilized and lived in. Rural life is dynamic, not static. Its essential, unchanging components are the least significant in terms of historical understanding. The appreciation of the *difference* between various rural configurations is a more fruitful line of inquiry. When people chose, at a given point in history, to congregate in towns, this brought about a radical change in the way they interacted with their human and physical environment. The villages that emerged alongside the towns were different in kind from earlier or later villages, as were those that came into being after urban collapse. The shifts in settlement evident in the Hula Valley and throughout the country were the consequences, intended or unintended, of purposive human action in specific historical circumstances, and the result is a dynamic, unrepetitive unfolding of rural and urban life. Inscribed in the landscape and imprinted in collective memory, this palimpsest of shifting social forms and interrelations was to have a significant impact on the history of the region, and indeed contribute to the diversity that engendered the unique cultures of Canaan and ancient Israel.

Notes

1 These are described in detail in Greenberg 1996b, from which the following paragraphs are abstracted. For preliminary descriptions see Greenberg 1990; 1996a. The full justification of the settlement map for each period will be presented in a forthcoming publication.

2 "Every established order tends to produce (to very different degrees and with very different means) the naturalization of its own arbitrariness" (Bourdieu 1977:164).

REFERENCES

Adams, R.M. and H.J. Nissen
1972 *The Uruk Countryside*. Chicago: University of Chicago Press.

Amiran, R.
1961 Tombs of the Middle Bronze Age I at Ma'ayan Barukh. *'Atiqot* (ES) 3:84-92.

Bahat, D.
1972 The Date of the Dolmens near Kibbutz Shamir. *Israel Exploration Journal* 22:44-46.

Biran, A.
1994 *Biblical Dan*. Jerusalem: Israel Exploration Society.

Beit-Arieh, I.
1981 A Pattern of Settlement in Southern Sinai and Southern Canaan in the Third Millennium B.C. *Bulletin of the American Schools of Oriental Research* 243:31–55.

Ben-Tor, A.
1997 Introduction. Pp. 1–13 in A. Ben-Tor and R. Bonfil, eds. *Hazor* V. Jerusalem: Israel Exploration Society.
1998 Tel Hazor, 1998. *Israel Exploration Journal* 48:274–278.

Blouet, B.
1972 Factors Influencing the Evolution of Settlement Patterns. Pp. 149–172 in P.J. Ucko, R. Tringham, and G.W. Dimbleby, eds. *Man, Settlement and Urbanism*. London: Duckworth.

Bourdieu, P.
1977 *Outline of a Theory of Practice*. Cambridge: Cambridge University Press.

Braun, E.
1997 *Yiftah'el, Salvage and Rescue Excavations at a Prehistoric Village in Lower Galilee, Israel* (IAA Reports 2). Jerusalem: Israel Antiquities Authority.

Covello-Paran, K.
1996 Middle Bronze Age Burial Caves at Hagosherim, Upper Galilee. *'Atiqot* 30:71–83.

Dever, W.G.
1980 New Vistas on the EB IV ("MB I") Horizon in Syria-Palestine. *Bulletin of the American Schools of Oriental Research* 237:35-64.
1986 'Abel-Beth-Ma`acah: "Northern Gateway of Ancient Israel". Pp. 207–222 in L.T. Geraty and L.G. Herr, eds. *The Archaeology of Jordan and other Studies, Presented to Siegfried H. Horn*. Berrien Springs MI: Andrews University Press.
1989 The Collapse of the Early Bronze Age in Palestine—Toward a Systemic Analysis. Pp. 225–246 in P. de Miroschedji, ed. *L'Urbanisation de la Palestine à l'âge du Bronze ancien* (British Archaeological Reports, International Series 527). Oxford.
1995 Social Structure in the Early Bronze Age IV Period in Palestine. Pp. 282–296 in T.E. Levy ed. *The Archaeology of Society in the Holy Land*. London: Leicester University Press.

Eisenberg, E.
1985 A Burial Cave of the Early Bronze Age IV (MB I) near 'Enan. *'Atiqot* (ES) 17:59–74.
1989 The Chalcolithic and Early Bronze Occupations at Tel Teo. Pp. 29–40 in P. de Miroschedji ed. *L'Urbanisation de la Palestine à lâge du Bronze ancien* (British Archaeological Reports, International Series 527). Oxford.

Eisenberg, E., A. Gopher, and R. Greenberg
2001 *Tel Te'o—A Neolithic, Chalcolithic, and Early Bronze Age Site in the Hula Valley* (IAA Reports). Jerusalem: IAA.

Eisenstadt, S.N.
1988 Beyond Collapse. Pp. 236–243 in N. Yoffee and G.L. Cowgill eds. *The Collapse of Ancient States and Civilizations*. Tucson: University of Arizona Press.

Epstein, C.
1974 Middle Bronze Age Tombs at Kefar Szold and Ginosar. *'Atiqot* (HS) 7:13–39 (Hebrew, with English summary on pp. 2*–6*).

Epstein, C.
1985 Dolmens Excavated in the Golan. *'Atiqot (*ES*)* 17:20-58.

Falconer, S.E.
1994 The Development and Decline of Bronze Age Civilization in the Southern Levant: A Reassessment of Urbanism and Ruralism. In C. Mathers and S. Stoddart eds. *Development and Decline in the Mediterranean Bronze Age* (Sheffield Archaeological Monographs 8). Sheffield: Collis. Pp. 305–333.

Finkelstein, I.
1991 The Central Hill Country in the Intermediate Bronze Age. *Israel Exploration Journal* 41:19-45.
1995 Two Notes on Early Bronze Age Urbanization and Urbanism. *Tel Aviv* 22:47-69.

Finkelstein, I. and D. Ussishkin
2000 Archaeological and Historical Conclusions. Pp. 576–605 in I. Finkelstein, D. Ussishkin, and B. Halpern eds. *Megiddo III: The 1992–1996 Seasons*. Tel Aviv: Yass Publications in Archaeology.

Genz, H.
2000 Grain-Wash Decoration in Early Bronze Age III? The Evidence from Khirbet ez-Zeraqon. Pp. 279–286 in G. Philip and D.W. Baird, eds. *Ceramics and Change in the Early Bronze Age of the Southern Levant*. Sheffield: Sheffield Academic Press.

Gophna, R.
1989 From Village to Town in the Lod Valley – A Case Study. Pp. 97–108 in P. de Miroschedji ed. *L'Urbanisation de la Palestine à l'âge du Bronze ancien* (British Archaeological Reports, International Series 527). Oxford.

Greenberg, R.
1990 The Settlement of the Hula Valley in the Urban Phase of the Early Bronze Age. *Eretz Israel* 21: 127–131 (Hebrew, with English summary on pp.105*–106*).
1996a The Early Bronze Age Levels. Pp. 85–160 in A. Biran, D. Ilan, and R. Greenberg *Dan* I. Jerusalem: Hebrew Union College.
1996b *The Hula Valley from the Beginning of the Early Bronze Age to the End of the Middle Bronze Age IIA: A Study in Regional Archaeology*. Ph.D. dissertation, Hebrew University, Jerusalem (Hebrew, with English abstract).

1997 Area A: The Early Bronze Age; The Early Bronze Age Phases in Area L. Pp. 17–24; 183–193 in A. Ben-Tor and R. Bonfil, eds. *Hazor* V. Jerusalem: Israel Exploration Society.

1999 Uniformity as Ideology in the Early Bronze Age II. Pp. 15–30 in A. Faust and A. Maeir eds. *Material Culture, Society and Ideology: New Directions in the Archaeology of the Land of Israel.* Conference Proceedings, Bar-Ilan University, June 3 1999. Ramat Gan. (Hebrew, with English summary).

2000 Changes in Ceramic Production Between EB II and EB III in North Israel, Based on the Pottery of Tel Hazor and Tel Dan. Pp. 183–200 in G. Philip and D.W. Baird, eds. *Ceramics and Change in the Early Bronze Age of the Southern Levant.* Sheffield: Sheffield Academic Press.

2001 An Early Bronze Age Tomb at Gadot. *'Atiqot* 42.

Greenberg, R. and E. Eisenberg,

In press Egypt, Bet Yerah, and Early Canaanite Urbanization. In T.E. Levy and E.C.M. van den Brink, eds. *Egyptian-Canaanite Interaction During the 4th Through Early 3rd Millennia BCE* (New Approaches in Anthropological Archaeology). London: Leicester University Press.

Greenberg, R. and N. Porat

1996 A Third Millennium Levantine Pottery Production Center: Typology, Petrography and Provenance of the Metallic Ware of Northern Israel and Adjacent Regions. *Bulletin of the American Schools of Oriental Research* 301:1–24.

Greenberg, R., L.K. Horwitz, O. Lernau, H.K. Mienis, H. Khalaily, and O. Marder

1998 A Sounding at Tel Na'ama in the Hula Valley. *'Atiqot* 35:9–35.

Haggett P., A.D. Cliff and A.E. Frey

1977 *Locational Analysis in Human Geography.* London: E. Arnold.

Harrison, T.P.

1997 Shifting Patterns of Settlement in the Highlands of Central Jordan during the Early Bronze Age. *Bulletin of the American Schools of Oriental Research* 306:1–37.

Herzog, Z.

1997 *Archaeology of the City.* Tel Aviv: Yass Archaeology Press.

Hirth, K.G.

1978 Interregional Trade and the Formation of Prehistoric Gateway Communities. *American Antiquity* 43: 35–45.

Ilan, D.

1992 A Middle Bronze Age Offering Deposit from Tel Dan. *Tel Aviv* 19:247–266.

1996 The Middle Bronze Age Tombs. Pp. 163–329 in A. Biran, D. Ilan, and R. Greenberg. *Dan* I. Jerusalem: Hebrew Union College.

Joffe, A.H.

1991 Early Bronze I and the Evolution of Social Complexity in the Southern Levant. *Journal of Mediterranean Archaeology* 4:3-58.

1993 *Settlement and Society in the Early Bronze Age I and II, Southern Levant* (Monographs in Mediterranean Archaeology 4). Sheffield: Sheffield Academic Press.

Kamlah, J.

2000 Early Bronze Age Grain-Wash Decoration from Northern Jordan. The Evidence of the Zeraqon Survey. Pp. 287–294 in G. Philip and D. W. Baird, eds. *Ceramics and Change in the Early Bronze Age of the Southern Levant.* Sheffield: Sheffield Academic Press.

Kantor, H.J.

1992 The Relative Chronology of Egypt and Its Foreign Correlations before the First Intermediate Period. Pp. 3–21 in R.W. Ehrich, ed. *Chronologies in Old World Archaeology.* Chicago: University of Chicago Press.

Maeir, A.

1997 Tomb 1181: A Multiple-Interment Burial Cave of the Transitional Middle Bronze Age IIA–B. Pp. 245–346 in A. Ben-Tor and R. Bonfil eds. *Hazor* V. Jerusalem: IES.

Marfoe, L.

1979 The Integrative Transformation: Patterns of Socio-Political Organization in Southern Syria. *Bulletin of the American Schools of Oriental Research* 234:1-42.

Millon, R.

1988 The Last Years of Teotihuacan Dominance. Pp. 102–164 in N. Yoffee and G.L. Cowgill, eds. *The Collapse of Ancient States and Civilizations.* Tucson: University of Arizona Press.

Miroschedji, P. de

1989 Le processus d'urbanisation en Palestine au Bronze ancien: chronologie et rythmes. Pp. 63–79 in *idem.* ed. *L'Urbanisation de la Palestine à l'âge du Bronze ancien* (British Archaeological Reports, International Series 527). Oxford: BAR.

Naaman, N.

1982 The Land of Israel in the Period of the Egyptian Middle Kingdom. Pp. 136–161 in I. Ephal, ed. *The History of Eretz Israel: Introductions, The Early Periods.* Jerusalem: Yad Ben Zvi. (Hebrew).

Palumbo, G.

1991 *The Early Bronze Age IV in the Southern Levant: Settlement Patterns, Economy, and Material Culture of a 'Dark Age'* (Contributi e Materiali di Archeologia Orientale 3). Rome: University of Rome.

Portugali, J.
1989 Demographic Speculations on Evolution, Settlement, and Urbanism. *Archaeologia* (Bulletin of the Israel Association of Archaeologists) 2:8–18 (Hebrew).
1994 Theoretical Speculations on the Transition from Nomadism to Monarchy. Pp. 203–217 in I. Finkelstein and N. Na'aman, eds. *From Nomadism to Monarchy*. Jerusalem: Israel Exploration Society.

Portugali, J. and R. Gophna
1993 Crisis, Progress and Urbanization: The Transition from Early Bronze I to Early Bronze II in Palestine. *Tel Aviv* 20:164-186.

Postgate, J.N.
1994 *Early Mesopotamia: Society and Economy at the Dawn of History*. London: Routledge.

Richard, S.
1987 The Early Bronze Age. The Rise and Collapse of Urbanism. *Biblical Archaeologist* 50:22-43.

Rosen, B.
1994 Subsistence Economy in Iron Age I. Pp. 339–351 in I. Finkelstein and N. Na'aman, eds. *From Nomadism to Monarchy*. Jerusalem: Yad Ben Zvi.

Rowton, M.B.
1976 Dimorphic Structure and Topology. *Oriens Antiquus* 15:17-31.

Saidah, R.
1979 Fouilles de Sidon-Dakerman: l'agglomération chalcolithique. *Berytus* 27:29–55.

Schaub, T.
1982 The Origin of the Early Bronze Age Walled Town Culture of Jordan. *Studies in the History and Archaeology of Jordan* 1: 67–75.

Shaked, I.
1994 Qiryat Shemona, Giv'at Shehumit. *Excavations and Surveys in Israel* 12: 3–4.
1995 Qiryat Shemona, Giv'at Shehumit. *Excavations and Surveys in Israel* 14: 7–9.
1999 Qiryat Shemona, Giv'at Shehumit. *Hadashot Arkheologiyot/Excavations and Surveys in Israel* 109: 4–5.

Shaked, I. And Marder O.
1998 Survey of the Map of Metulla—Prehistoric Sites. *Excavations and Surveys in Israel* 18:4–5.

Stepansky, Y.
1992 Rosh Pinna Map, Survey. *Excavations and Surveys in Israel* 10:66–68.
1994 Rosh Pinna Map, Survey—1991. *Excavations and Surveys in Israel* 12:6–8.
1999 *The Periphery of Hazor during the Bronze Age, the Iron Age and the Persian Period: A Regional-Archaeological Study*. M.A. thesis, Tel Aviv University (Hebrew, with English summary).

Stech, T., J.D. Muhly and R. Maddin
1985 Metallurgical Studies on Artifacts from the Tomb near 'Enan. *'Atiqot* (ES) 17:75-82.

Wright, G.E.
1958 The Problem of the Transition Between the Chalcolithic and Early Bronze Age. *Eretz Israel* 5:37*-45*.

Yadin, Y.
1972 *Hazor* (The Schweich Lectures). London: Oxford University Press.

Yeivin, Z.
1966 Ancient Tombs at Kibbutz Hagosherim. *Bulletin of the Israel Exploration Society* 30:98–109 (Hebrew).

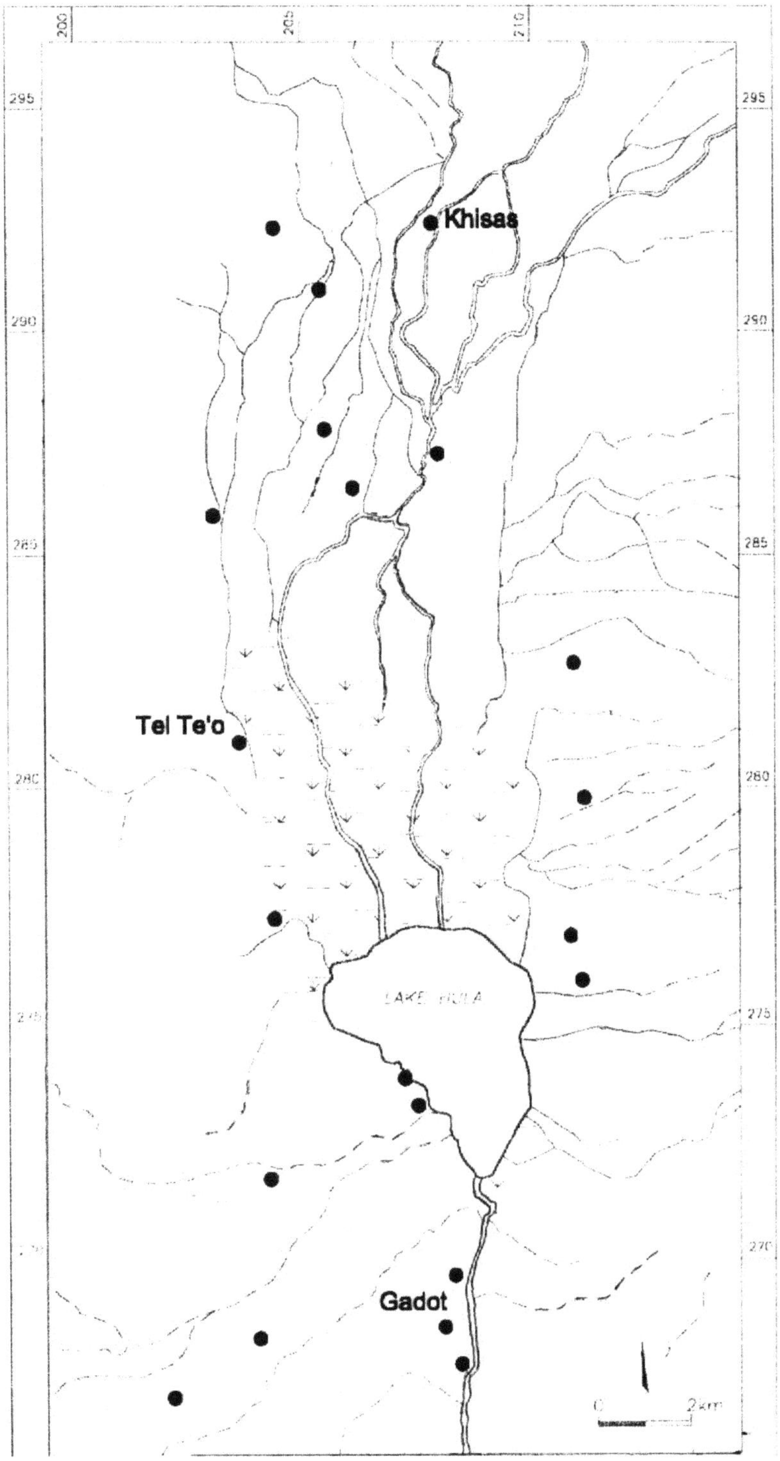

Figure 2-1: Hula Valley Settlement in EB I

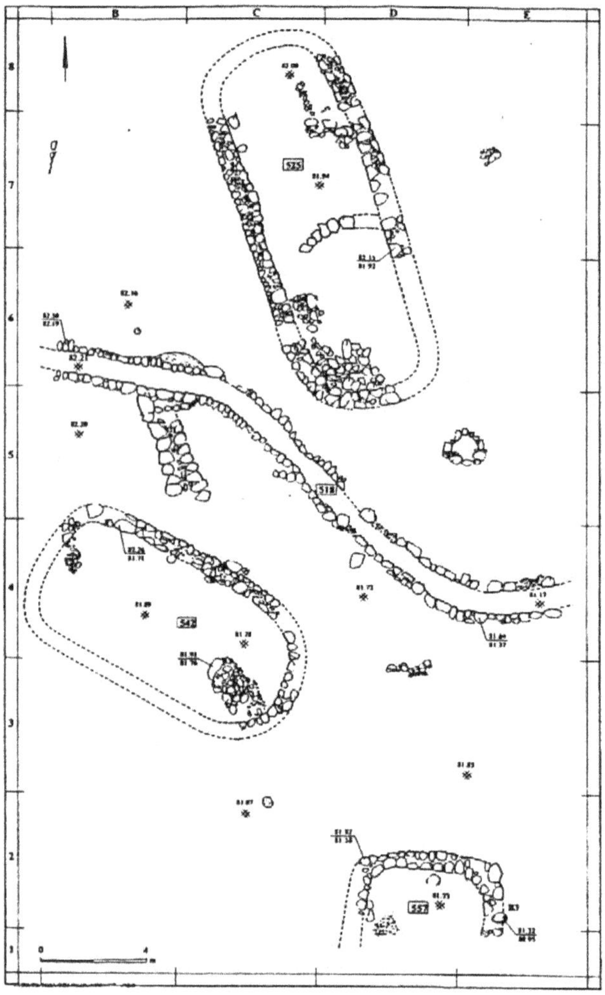

Figure 2-2: EB IA structures at Tel Te'o

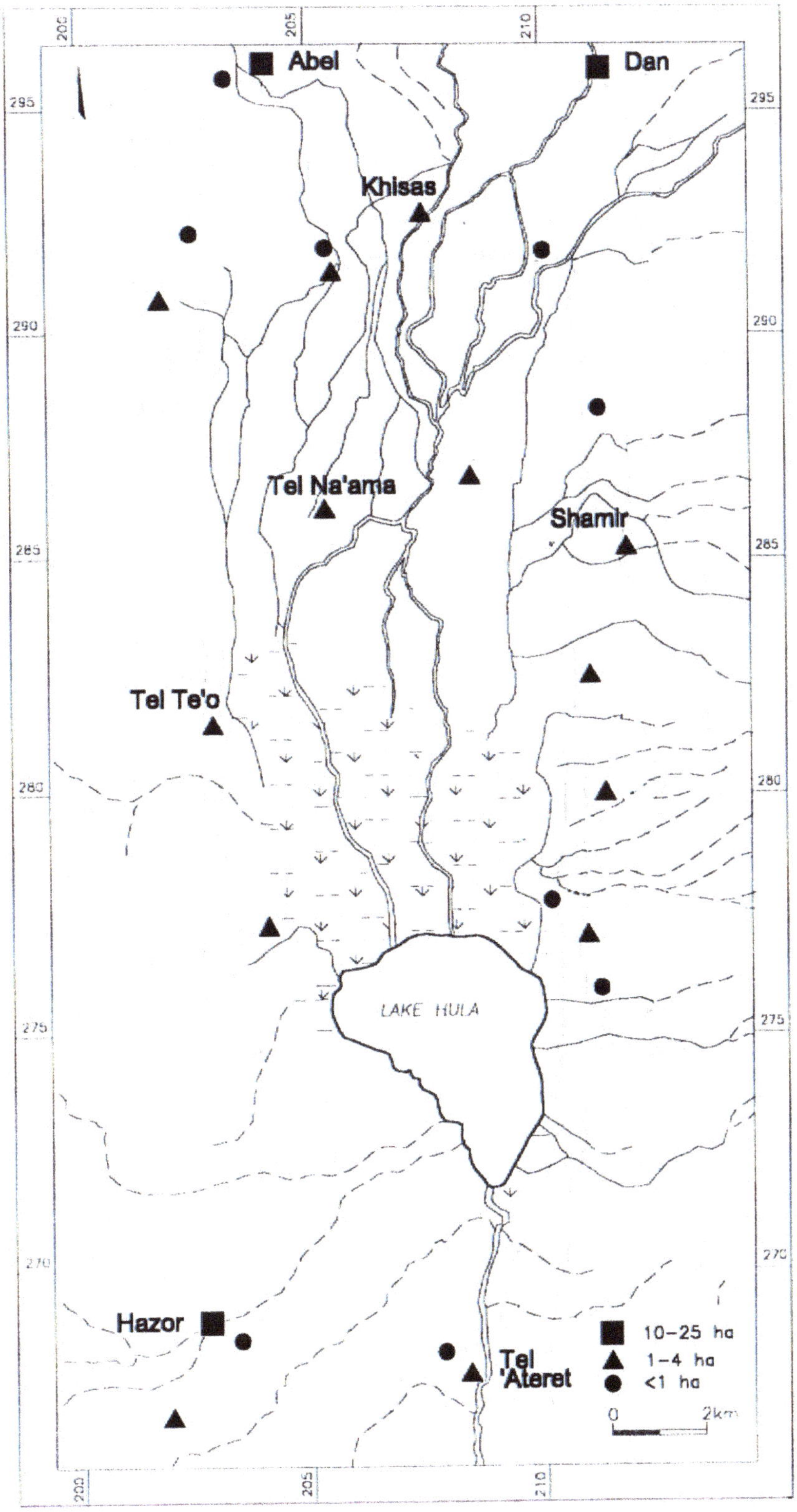

Figure 2-3: Hula Valley Settlement in EB II

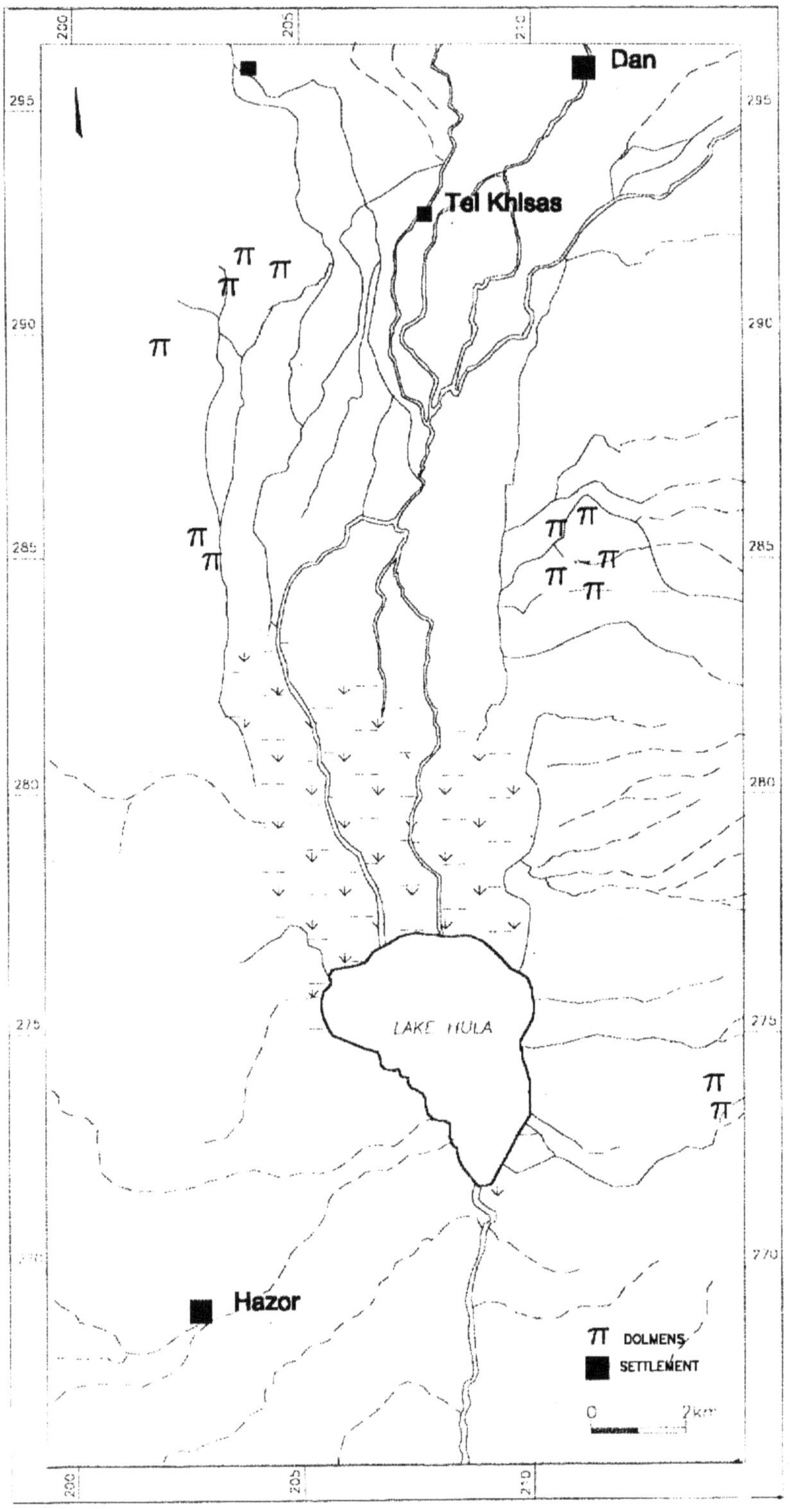

Figure 2-4: Hula Valley Settlement in EB III

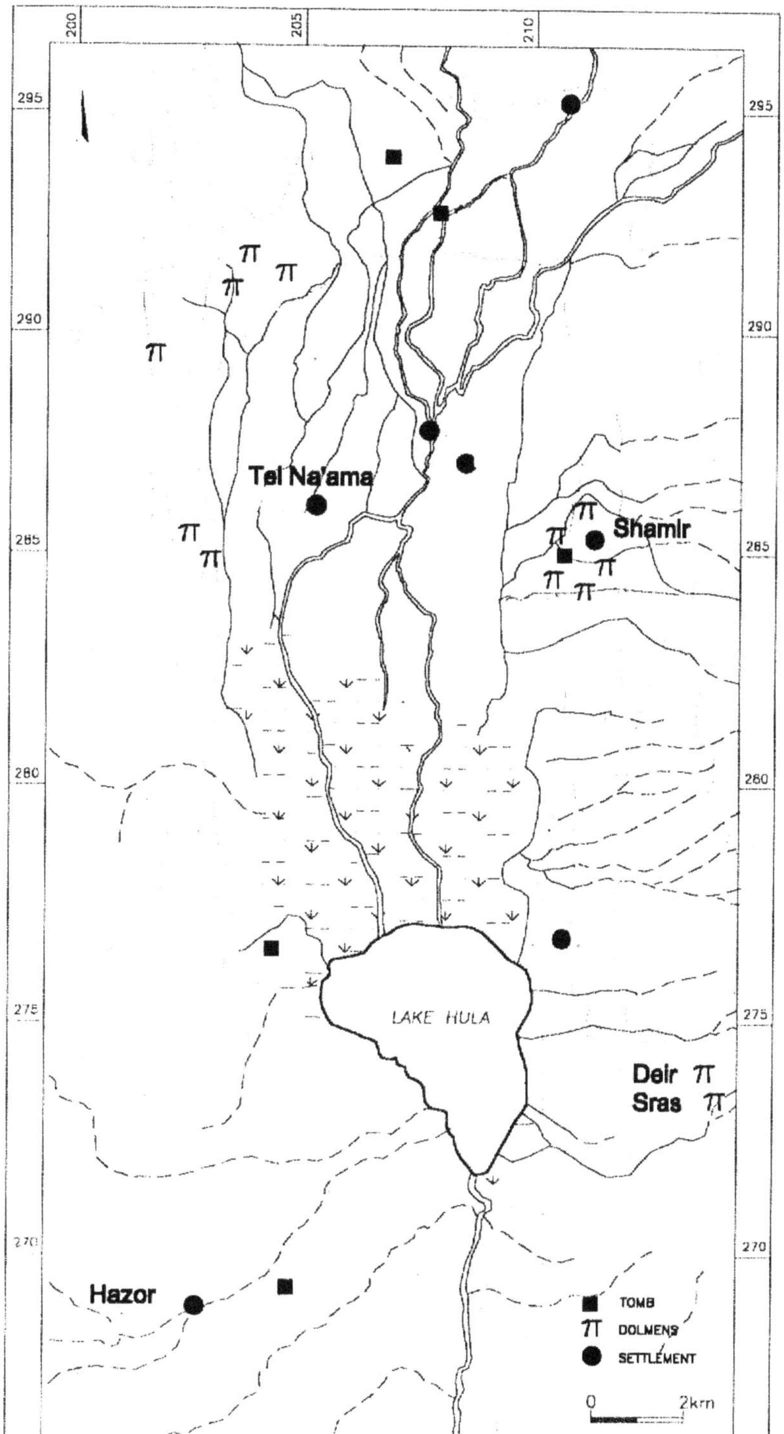

Figure 2-5: Hula Valley Settlement in the IBA

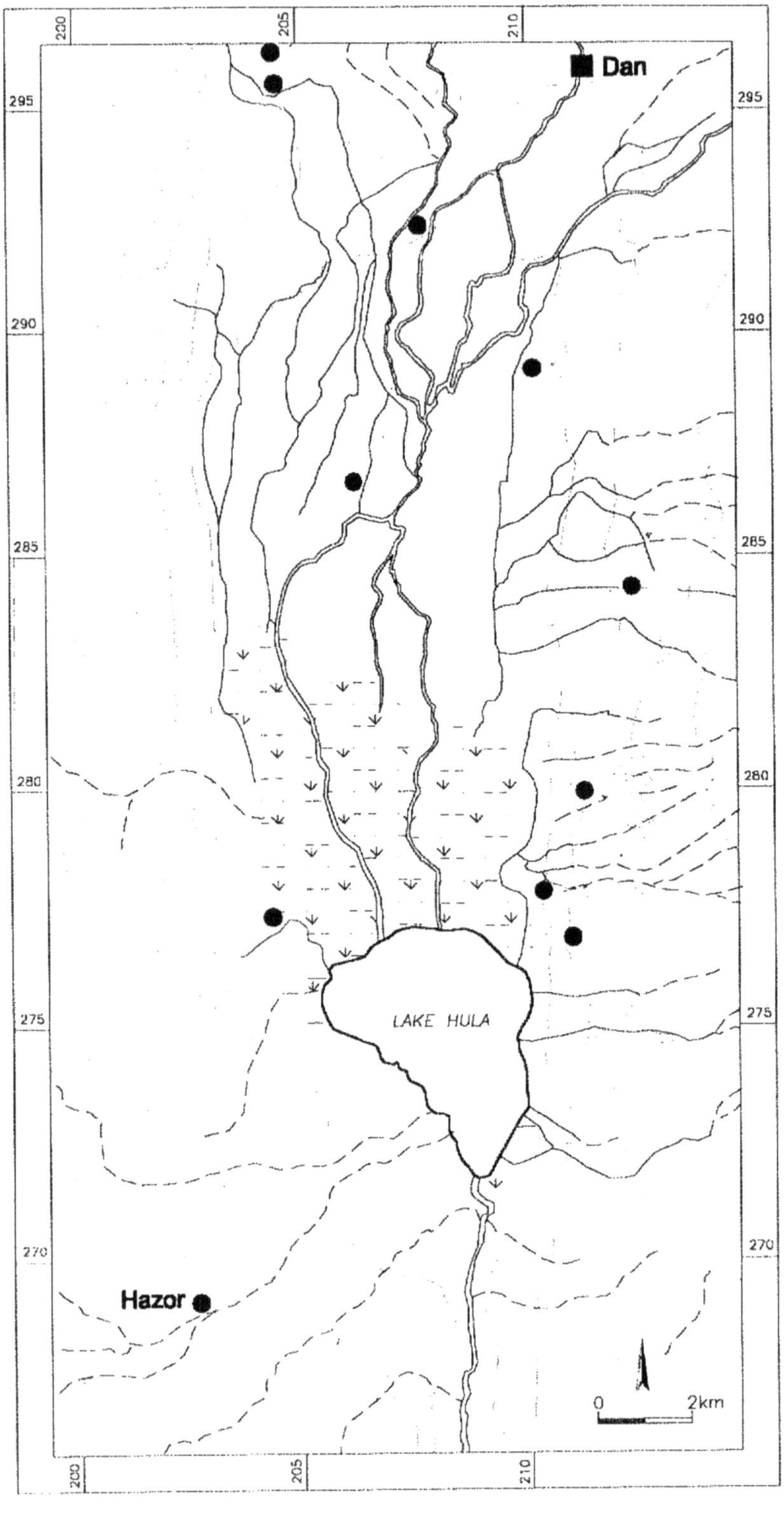

Figure 2-6: Hula Valley Settlement in MB I

The Rural Landscape of Palestine in the Early Bronze IV Period

WILLIAM G. DEVER

The Department of Near Eastern Studies
University of Arizona
Tucson, AZ 85721
U.S.A

The socio-eonomic structure of the Early Bronze IV period has been intensively discussed for several decades. Although in the past I have espoused the "Pastoral Nomadic Model" for the understanding of this period, in more recent years, critics of this model, and myself, have preferred explanations based on more flexible models, models in which the oscillating nature and relationships between the urban and rural parts of society are stressed. Based on this current understanding, the ideological reasoning behind the landscape changes during this period is studied. Utilizing methods of the "cognitive archaeology", an attempt is made to recreate the "mental map" of the population of Palestine during this period.

The papers in this volume were invited to focus on the "landscape" of ancient Palestine. For purposes of discussion, I shall define "landscape" here *archaeologically* as "the natural scenery and physical environment of a particular geographical region, together with the picture derived from any discernible human (i.e., cultural) impact that may have transformed the original setting." In the following pages, I shall use the Early Bronze IV period in Palestine and Transjordan (sometimes "Middle Bronze I" or "Intermediate Bronze"), *ca.* 2300-2000 BCE, as a case-study in landscape archaeology, raising two questions.[1] (1) What did Palestine actually *look* like in this period? (2) What environmental and/or cultural factors may be induced to *explain* the changes that took place, and consequently set off this enigmatic "non-urban" interlude from the great Early Bronze and Middle Bronze Age civilizations that preceded and followed it?

The Development of Landscape Archaeology

Although what may be called "landscape archaeology" has not yet developed into a well-defined and specialized sub-branch of archaeology, the impetus for it goes back to several earlier theoretical trends. These would include "settlement archaeology," pioneered by Adams, Willey, Chang, and others in the 1960's, which began to turn attention from individual sites to larger patterns of settlement types and distribution (Trigger 1989b:279-288; Renfrew and Bahn 1991:61-1000). Then the ecological thrust of the New Archaeology of the 1970's focused on the macro-environmental context of archaeological phenomena, or the study of "sites in their setting (Trigger 1989b:303-323). The multi-disciplinary character of the New Archaeology also resulted in the retrieval of a much larger and more reliable body of environmental data, through collaboration with geologists, geomorphologists, climatologists, palynologists, paleo-botanists, and other natural scientists. The emphasis on regional studies, and especially the widespread development of intensive surface surveys, typical of the 1980's, further heightened our awareness of the larger archaeological landscape and its importance.[2] "Systems theory" models, now somewhat discredited, nevertheless had the advantage of demonstrating how complex and interrelated natural (or organic) systems are, and how human culture (as "adaptation") is partly conditioned by these natural systems, principally environmental, of course.[3] Also in the 1980's, archaeologists' belated discovery and half-hearted employment of models drawn from the *Annales* school pointed out the significance of long-term cultural evolution, deeply rooted in slowly changing environmental factors (Braudel's *structures*, the "deep swells", *la longue durée*).[4] The introduction of "collapse" models" in the 1990's highlighted the delicate symbiosis between natural and cultural systems, particularly evident in the "cycles of advance-abatement" so typical of the ancient Near East (below).[5] Finally, current "post-processualist" approaches have stressed archaeological "context," as well as long-term settlement-history, in a way that will, I think, contribute powerfully to the emergence of a self-conscious and well-defined "landscape archaeology," the need for which is already clear.[6]

The Early Bronze IV Period as An Instructive "Case Study"

No period in the long archaeological history of ancient Palestine and Transjordan (hereafter simply "Palestine") provides a more instructive example of major shifts in settlement patterns--and thus of changing relations with the landscape--than the Early Bronze IV period, *ca.* 2300-2000 BCE (hereafter "EB IV"). Long a poorly documented and supposed "dark age," this period has been richly illuminated by recent discoveries. The current state of our knowledge is summarized by Gaetano Palumbo's magisterial *The Early Bronze Age IV in the Southern Levant. Settlement Patterns, Economy, and Material Culture of a "Dark Age"* (1991), so there is no need here to present the basic data.

In any case, all authorities today agree that EB IV characterizes a relatively brief, disjunctive, unique "episode" in Palestine's long settlement history, in which the cumulative urban occupation of the Early Bronze I-III era (*ca.* 3400-2300 BCE) comes to a complete end. It is then replaced for at least three centuries by a completely non-urban pattern of settlement, consisting mostly of hundreds of small, seasonal pastoral-nomadic encampments--many located in the marginal, semi-arid hinterlands of the Central Negev, the Jordan Valley, and southern Transjordan. There are only a relatively few more permanent agricultural or agro-pastoral villages, either remaining from EB or newly-founded in the heartland of Palestine.

The drastic displacement and relocation of the population in EB IV can best be documented simply by observing that to date not a single excavated site in either western or eastern Palestine exhibits an unbroken occupational sequence extending from EB III through EB IV and on into the urban renascence that takes place in the first phase of the Middle Bronze Age (after *ca.* 2000/1950 BCE; traditionally "Middle Bronze IIA," but now often, and more appropriately, "Middle Bronze I").[7] Nor do we have a single EB IV archaeological site that can be properly characterized as truly "urban" by any reasonable criteria, such as the presence of city walls, or monumental architecture of any kind--not to mention evidence for industry, international trade, stratified socio-economic structure, centralized political organization, or other typically "urban" characteristics.

The EB IV shift in settlement patterns, unique in Palestine's long archaeological history, radically transforms the landscape. This phenomenon has long been recognized, but it has been difficult to explain by any available theoretical models. Long ago Albright, Glueck, Kenyon, and others thought that the hiatus in urban settlements was due to "Amorite" invasions and destructions, i.e., incursions of West Semitic pastoral nomads well known to us from late 3rd-early 2nd millennium BCE texts from Mesopotamia.[8] In the 1960's Paul Lapp (1970) and Moshe Kochavi (1967) invoked "Kurgan" newcomers from the Caucasus region to account for the widespread disruptions in Palestine. In the early 1970's (Dever 1971;1973), I proposed "pastoral nomadism" as an overall model to account for several features of EB IV that were clear by then. These included: (1) the lack of any real EB II-III "destruction levels"; (2) the heavy concentration of EB IV sites in the semi-arid regions more suitable for pastoralism; (3) the ephemeral, apparently seasonal, and often short-lived occupation at many typical EB IV sites; (4) the strongly "regional" cast of the EB IV material culture, so that Amiran had already distinguished regional ceramic "families", and I later expanded this notion into larger "geographico-cultural "assemblages. These groupings became my Families "North"; "North Central"; "Central Hills;" "Jordan Valley"; and "Southern", later augmented as new evidence became available by a Family "TR" in Transjordan (Dever 1973; 1980).

As the excavation of EB IV sites expanded, "regional families" served rather well as an explanatory model, even to the extent of predicting, for instance, what ceramic types would likely be found in a given area. If predictability is the test of a model, this one "worked" for nearly 20 years, and to a large degree it still does. By the late 1980's, however, my "pastoral nomadic" model was coming under increasing criticism as new data came to light.[9]

In western Palestine, more extensive and deliberate surveys and excavations, while revealing many more Negev encampments, also turned up a number of apparently permanent agricultural villages in such areas as the central hill country and at Sha'ar ha-Golan in the upper Jordan Valley. In Transjordan, the upper Jordan Valley also yielded village sites, like Tell el-Hayyat and nearby Tell Abu en-Niaj.[10] And in southern Jordan, the large mound of Kh. Iskander on the Wadi Wala was hailed as the first "urban" EB IV site to be discovered, complete with a massive city wall. It should be noted, however, that Tell el-Hayyat has clear EB IV sherds in the basal layer, but absolutely no architecture. And the "EB IV city wall" at Kh. Iskander turns out to be no more than an ephemeral use (if any) of an EB II town wall, as I suspected all along.[11] Thus, I think that it is still true that we do not possess a single EB IV site that can be called "urban" by the usually accepted criteria. Where does that leave us?

"Pastoral Nomadism" and Its Critics

Critiques of my "pastoral nomadic" model (above) have tended in my judgment to overemphasize the relatively few EB IV village sites at the expense of the still preponderant campsites, together with the continued discovery of vast isolated shaft-tomb cemeteries with disarticulated, secondary burials--the latter an *undoubted* attestation of pastoral nomad lifestyles. Critics have also overlooked (1) the fact that my "pastoral nomadic" model was a product of the 1970's, when most EB IV sites that we had were located in the Negev and other marginal zones and were indeed nomadic camps; and (2) the fact that I never advocated the "pastoral nomadic" model for all of Palestine and Transjordan.[12]

Despite these caveats, which I hope are not overly defensive, I took critics seriously; and by 1995 I had responded with a modification, i.e., a model that embraced a socio-economic structure of mixed agro-pastoralism and "ruralism" (Dever 1995). Aware that this is more a description of the EB IV "landscape" than a full-scale and properly articulated explanatory model, I have left matters there. Meanwhile, both my critics and I have tentatively advanced notions of more nuanced possibilities of southern Levantine societies that may be positioned along a *continuum* from fully nomadic to fully sedentary. The notion of a culture in EB IV that is largely "rural" could easily be accommodated by such a model.

Before proceeding further, however, we need to define what would appear to be basic aspects of the EB IV society that any model must accommodate: (1) "Pastoral nomadism" may be defined as "a specialized offshoot of sedentary agriculture, which produces a society and economy characterized by dependence primarily on the herding of animals, temporary dwellings, and seasonal movements." (2) "Mixed agro-pastoralism" may be defined as "a sedentary or semi-sedentary society and economy characterized by village-based agriculture, combined with limited seasonal movements and herding of animals."[13]

The more recent theoretical emphasis overall is thus on more flexible models, stressing dynamic and cyclical *oscillations* between what might be termed "the rise and collapse of complex societies"; "advance/ abatement"; "integration disintegration"; "centripetal and centrifugal forces"; or "specialization/despecialization."[14] In that case, the EB IV horizon is clearly one of urban collapse, decentralization of population, and despecialization in subsistence systems, as well as a period of more "egalitarian" social structure. The EB IV archaeological landscape, I would argue, dramatically reflects all these shifts along the continuum posited above. Yet how can we *account* for these changes?

Toward Better Explanatory Models? "Post-processual" and "Cognitive" Archaeology

None of the above paradigms really constitutes an adequate explanatory model, as the latter is understood in wider archaeological and anthropological circles. Old systems-theory approaches still appeal to me, despite the fact that they are now considered by many to be inadequate and counter-productive. At least systems-theory stresses the *formative role* of the environment in cultural change, or our "archaeological landscape" here. But can we get further in our search for causal explanations?

It may be that today's climate of post-processual or "contextual/cognitive" archaeology offers some hope. Post-processualism began as early as the mid-1970's, largely as a reaction against the functionalist, determinist, and reductionist excesses of 1960's-1970's-style New Archaeology--especially its anti-historical bias. To judge from the most recent literature on theory in archaeology, "post-processualism" is now the reigning paradigm in most circles in both Europe and North America. The post-processualist "movement," however, if it may be so termed, is not by any means homogeneous, but is intentionally eclectic; it even embraces some of processualist archaeology's features, such as emphasis on research design and the multi-disciplinary thrust.[15]

The newer approach, however, is easily distinguished by the terms it prefers, i.e., not only "post-processual," but also "contextual," "cognitive," "interpretive," and "new critical." As for the alternate terms, Ian Hodder, one of the initiators of post-processualist thought already in the early 1980's, has pointed out that "context"--which all archaeologists recognize as being of crucial importance in archaeological interpretation--can be construed as "context," i.e., read "*with* texts," so as to contribute to history-writing. Hodder and a number of other British archaeologists have gone further, to suggest that archaeological data can even be "read *as* text." That is, artifacts can be interpreted in very much the same way as texts, as "encoded messages" from the past. One implication is that once we learn the "language" of material cultural remains--the vocabulary, grammar, and syntax--we can employ hermeneutical principles in their interpretation that are similar to those for textual remains. Another and more significant implication, however, is that archaeological evidence has the potential of becoming primary data for history-writing--equal to or even superior to ancient texts as valid information about the past.[16] In that case, the abundant data that we now possess might help us finally write a *real* "cultural history" of the EB IV period in Palestine. Even though we have no texts for this period (but see below), an artifactually and environmentally based reconstruction of the period might at least lay the foundations for a socio-economic, political, cultural, and technological history, if not a "history of ideas."

This leads to a discussion of a related term now in vogue, "cognitive archaeology." This implies that precisely what the New Archaeologists argued was impossible may be possible after all, that is, to "read" from artifacts not only behavior, but something of the *thought* that produced that behavior. As Hodder has expressed it, "history in the sense intended here involves also getting at the *inside* of events, at the intentions and thoughts of subjective actors" (1986:77). Elsewhere he quotes Colingwood approvingly, to the effect that "to study history is to try to get at purpose and thought" (1986:91).

Using the Newer Paradigms to Reconstruct an "EB IV Landscape"

Several more recent discussions of post-processual theory, such as the recent edited volume of essays, *Reader in Archaeological Theory: Post-processual and Cognitive Approaches* (Whitley 1998), have utilized the concept of "cognition" as a goal of archaeology in a way that may help to elucidate a concept such as "archaeological landscape," which can admittedly be vague. Persis B. Clarkson's paper in the latter volume, "Archaeological Imaginings: Conceptualization of Images," addresses directly our question here: What *is* a "landscape?" Clarkson quotes Ingold that "landscape is a cultural process," then goes on to specify that culture is not only a framework for perceiving the natural world, but also for interpreting it to oneself and to others (1998:120, 121)--and, I would add, for *transforming* that world through the cultural process. Hodder himself had observed that "'space' is converted into '*place*,' defined as a center of meaning or a focus of human emotional attachment" (1987:140, italics WGD). Similar concepts of "landscape" have long been employed by geographers; and there is even a branch of historical geography sometimes known as "environmental perception."[17]

What I shall attempt here is not another map of known EB IV sites,[18] but rather a cognitive or "conceptual map" such as the EB IV people themselves might have had. I shall then use that mental template in an attempt to *explain* the unique settlement pattern of that period--the "archaeological landscape"--in both its natural and cultural dimensions.

Even a few years ago such an attempt would have been dismissed contemptuously as entirely speculative; one recalls Binford's remark that "archaeologists are poorly equipped to be paleo-psychologists" (Binford 1972:198). In this vein, virtually all previous discussions of pastoral nomadism in the Levant have taken "ecological compulsion" to be the *only* factor in determining the lifestyle of pastoral peoples, as well as describing it.[19] Thus Johnson's much-quoted work *The Nature of Nomadism* (1969) takes "ecological compulsions and regular migrations" as its starting point in attempting to develop an all-embracing definition of pastoral nomadism. Despite the usefulness of his study, Johnson alludes to "cultural" factors only once, and then only in the old-fashioned sense of "culture areas." Yet this narrowly materialist, functionalist, and determinist ecological approach was typical of the New Archaeological approach of the 1970's, and it has continued even in some of the most recent archaeological and anthropological literature.

In today's climate of post-processual and interpretive archaeology, many attempts at "cognitive archaeology"--at such "mental mapping" as I envision--are being made. One thinks of Deetz' analysis of early New England houses and their furnishings; Glassie's similar study of folk-housing in Georgia; Muller's studies of the American southwest; Washburn's study of ceramic design; Hall's interpretation of Indian peace-pipes; Leroi-Gourham's elucidation of Paleolithic cave art; Clarkson's attempt at explaining the enigmatic geoglyphs in Nazea, Peru; David Whitley's exploration of the meaning of rock-drawings of the Late Prehistoric Coso culture in eastern California (up to *ca.* 1200 CE); and Dorothy Hoster's investigation of the specific symbolic meanings attached to metallurgy in Western Mexico *ca.* 600-1500 C.E.[20]

All of these and many other recent efforts of cognitive archaeologists aim wherever possible to get at "the *meaning* of things," which I find exhilarating after two decades of hearing New Archaeologists dismiss ideology as irrelevant in explaining cultural change.[21] Yet precisely *how* can we go about the attempt in the case of the EB IV landscape, without plunging into the pitfall of unbridled, subjective speculation of which "cognitive" archaeology has been accused by Binford and other unreconstructed processualists?[22]

The basic method employed in all the case-studies cited above rests largely on ethnographic observations and then employs arguments drawn from analogy. Following that lead, it would be obvious for us to look at the extensive literature on pastoral nomadism in the Levant from the Neolithic to modern times. Indeed, I have done so previously in attempting to understand EB IV settlement patterns in the semi-arid marginal zones (Dever 1973; 1980; 1992). Yet all such efforts have been confined largely to description of the phenomenon, rather than ideological explanations, as noted above. And the many *rural* villages now known in the heartland of ancient Palestine in EB IV have scarcely been included in the picture at all until now. A further difficulty is that the well-known cultural regionalism of Palestine in EB IV would seem to preclude any general explanatory theory for overall settlement patterns.

A recent comprehensive survey by Gophna (1992) has discussed regionalism, but in a way that seems to me to minimalize its significance. That intuition, however, even though it runs counter to my own earlier views, may be essentially correct, simply because we have so much new data that generalizations may now seem more difficult. If one analyzes Gophna's region-by-region summary carefully, however, it turns out that his data do support one significant generalization. In *all* regions--not just the semi-arid regions--most EB settlements are established *de novo*. Reoccupation of older *tell*-sites is the exception; and even in this case, the occupation is neither large, nucleated, nor in any way urban in character. It seems evident that the vast majority of the EB IV population lived in rural, even exceptionally remote areas, and had little sustained contact with any urban centers. For me, an intriguing question: *Why* did the EB IV folk by and large deliberately avoid the prime locations that had attracted intensive occupation throughout Early Bronze I-III (a period of at least a thousand years), i.e., those sites with optimum soil and water resources, as well as defensible positions? I do not think that the question has been previously raised in quite this way, but it may now be productive to do so. In short, is there at work in the EB IV shift in settlement patterns a *cultural concept* of the "ideal landscape"; and if so, what factors determined it?

A Pastoral Nomadic "Mentality"? Ethnographic Considerations

I cannot, of course, document here a sort of "paleo-Bedouin mystique" that could adequately explain the EB IV preference for dispersed, ephemeral, and often seasonal settlement sites. But enough is known of pastoral nomadic lifestyles from numerous ethnographic studies to suggest that there *are* several constants, i.e., cultural constraints that would tend to influence the concept of the ideal environment or "landscape."

(1) Pastoral nomads nearly always deliberately *prefer* a peripatetic existence, even under conditions of severe hardship; and they tend to look with contempt on town-dwellers and urbanites. The Bedouin regard themselves, whether accurate historically or not, as the "true Arabs" of the desert.[23]

(2) Pastoral nomads typically value their untrammeled freedom of movement over any supposed "advantages" that a settled lifestyle might offer them, and in particular they resist the attempt of the central authorities to settle them forcibly.

(3) Pastoral nomads take great pride in their economic independence and "eqalitarian," tribally-organized lifestyle, even when these tend to be more social fiction than fact.

In all the above, we see a modern notion of a "cultural landscape," which if it could actually be realized would idealize and favor what we may call the "countryside," or the rural component of society. It is possible, of course, to exaggerate this predisposition against urban life in the ancient and modern Middle East, positing an unremitting struggle between what Reifenberg called picturesquely "the Desert and the Sown"--a misconception that goes back at least as far as pre-modern Arab historians like Ibn Khaldun.

Ibn Khaldun (1332-1406 CE) was a Maghrebi philosopher who first formulated a view of the evolution of Arabic tribes and states that is still influential in Arab historiography today. His was in effect a logico-deductive model that Caton describes as "holistic in conception, complex, and subtle, embracing human psychology, social organization, the state, religion, and economy in accounting for human reality" (Caton 1990:89). For Ibn Khaldun, Arab "solidarity" is a central issue; and since this must be based on blood ties, the group that has the purest lineage, the best "pedigree," comprises the Bedouin. The more settled "hill Arabs" (or *fellahin*) have become effete, have lost their fierce, autonomous sense of family, of loyalty, and especially of *honor*. The latter never stems from coercion for Ibn Khaldun, but rather from *'assabiya*, "group feeling." "Honor," based on the solidarity of the group, is thus not only a collective ideology, i.e., cultural, but is essentially religious in nature. It is imbued with such rich symbolism and elaborate social rituals that it constitutes much of what I have called here a "mystique." It is one that is *especially* characteristic of pastoral nomads, is indeed regarded as being at the heart of a sort of "desert code." [24]

It is worth noting that this sense of "group solidarity" that is so typical of pastoral nomadic societies has persisted for many centuries, *despite* the fact that the transcendent concept of unity in Arab societies is the Islamic *umma*, or larger "community of believers"; and it is the settled urban society that is the civilized Islamic ideal.[25] I would argue that economic and political considerations alone cannot account for this persistence, only an ideology that no matter how much partly social construct nevertheless sustains an ideal and an idealized picture, a "cultural landscape."

A modern analysis of tribal societies that in some ways builds on Ibn Khaldun's notions of the opposition of tribe and state is that of Michael Rowton. In a series of articles he has developed the model of "enclosed nomadism," stressing that most Levantine societies have historically been "dimorphic."[26] The urban element and the rural-pastoral element are not, therefore, opposing "ethnic" groups, but constitute rather two aspects of the same society, representing differing and to some degree complimentary subsistence strategies and socio-political structures. The implication of Rowton's model is that most known Middle Eastern societies, past and present, are best comprehended as "hybrid" entities, situated somewhere along a *continuum* from a hypothetical fully nomadic to a fully sedentarized stage of cultural evolution. Thus the unique and significant factor in cultural change in Levantine societies lies in the characteristic cyclical *oscillation* between the two poles over long time-periods, first one then the other prevailing. If one adopts this model, the EB IV period represents a cycle in which the rural-pastoral morpheme predominates for several centuries, unusual only in that it almost entirely eclipses the urban morpheme. I would argue that it is precisely that shift that creates the unique EB IV "landscape" of small, dispersed settlements.

It is not, however, a simple or one-way influence; people shape the landscape, and at the same time, it shapes them and their way of life, in a dynamic, ever-changing symbiosis. In the case of EB IV, an equilibrium was created that lasted for several centuries, until the urban renascence at the beginning of the Middle Bronze Age, *ca.* 2000 BCE. That equilibrium, in my view, was not fortuitous but was often the result of deliberate *cultural choices*--of "ideology," if you will, which post-processual archaeologists now correctly recognize as no mere "epiphenomenon in culture change."[27] To put it another way, EB IV was not so much a period of "maladaptation," much less of widespread and total collapse, as it was one of *changing cultural values*. As observed earlier, "landscape is a cultural process." The EB IV landscape was *not* the result merely of "ecological compulsion," i.e., of necessity, but was also a matter of choice. And it was certainly not the result of the coercion of the urban authorities, for in EB IV Palestine there were none.

As I noted above, very few ethnographers working on the presumption of "ecological compulsion" have taken *ideology*, much less our "cognitive map" here, into consideration in seeking to describe pastoral nomadic societies. Fredrick Barth's great pioneering work *Nomads of South Persia: The Basseri Tribe and the Khamseh Confederacy* is a refreshing exception. Barth's Appendix 1 (1961:135-153), entitled "The Ritual Life of the Basseri," begins with the observation that outwardly the Basseri exhibit what appears to be an extreme poverty of ritual life. Certainly they are not formally "religious," in the sense of the life of urban Muslims. Barth observes shrewdly, however, that it is the *migrations themselves* that form a yearly cycle--a sort of "liturgical year," we would say--and that "it is terms of them that the average Basseri conceptualizes time and organizes his life" (1961:137). Barth then devotes some 15 pages to describing rituals inherent in pastoral nomadic lifestyles that serve to express the social unity of the group and its cultural values. The meanings and symbolic aspects of these activities, Barth says, "are of the same logical order and partly of the same form as many of the rituals of religious ceremonies. . . The *context* of these meanings is the cycle of migrations, which dominates the life and organizes most of the activities of the Basseri" (1961:148; italics WGD).

Arguing as I have here on the basis of ethnographic parallels for the existence of an EB IV "mental map" does not, of course, explain the *ultimate causes* for the dramatic shifts that are well documented in that period. People at that time may have initiated some of these changes deliberately, and in full or partial knowledge of the consequence, as a matter of cultural preference. But we still do not know *why* they did that in the first place, any more than we know what triggered the following urban renascence, the consequent disappearance of virtually all the EB IV settlements, and the making of a drastically new urban landscape, probably within a generation or two. I am arguing largely from ethnographic analogy, and that, however suggestive, has its limitations.

Some Textual Data

Post-processualist archaeology's attempt to get at cognition employs, as we have seen, not only the notion of reading material culture as "text," but also actual historical texts where available (Hodder's "context" as con-text, "with text"). It is well known that we have no texts whatsoever from Palestine in EB IV. We do possess, however, texts of the general period from both Egypt and Mesopotamia, some of them dealing in considerable detail with pastoral nomads who surely are directly comparable with the pastoral nomads of Palestine in EB IV--especially the well-known West Semitic "Amorites" of upper Mesopotamia and Syria. The Drehem texts of the Ur III period, *ca*. 2050-1950 BCE, are particularly instructive, as are the Mari Letters a century or two later.[28] Both provide an extensive description of pastoral nomads of the late 3rd/early 2nd millennium BCE., in which we see precisely the avoidance of urban lifestyles, and the resistance to forced sedentarization I have posited of pastoral nomads in Palestine in EB IV. The Mari texts even employ an Akkadian term that seems to reflect elements of the "cognitive map" that I have posited. The frequently used term *nawûm* is sometimes rendered simply "pastureland" or "encampment"; but Edzard, Rowton, Matthews, and others have pointed out that in larger context it means something like "the total pastoral group and its surroundings." This is a *collective concept* embodying the pastoralists themselves, their herds, and the entire camping area through which they migrate. The *nawûm* is thus rather like the Brahui form *Khalle* of the Khalot Khanate of West Pakistan, which denotes the mobile herding unit in its entirety--people, tents, encampments, and animals (Matthews 1978:58-63).

It is true that in Mesopotamia, Syria, and Palestine all these nomadic groups were eventually partly assimilated in the later, fully developed "Amorite" kingdoms and in the Middle Bronze Age of Palestine. But it is clear that for *centuries*, a "nomadic ideal" or mystique still prevailed. The Old Assyrian kinglist, naming later fully acculturated "Amorite" kings, begins with a proud reference to "17 kings who lived in tents," i.e., real or eponymous nomadic ancestors. Such nostalgia for the pastoral landscape and lifestyle is also familiar from the Hebrew Bible, where the ancestors of later Israel are portrayed in lavish detail as steppe-dwellers and migratory pastoralists, resistant to urban authorities.[29]

From roughly the same period in Egypt comes the well-known "Tale of Sinuhe," which I would argue can be taken as a contemporary, vivid, eye-witness description of actual conditions in the countryside of Syria-Palestine.[30] Note the following features of the story. (1) When Sinuhe succeeds in crossing the Sinai desert and entering southern Palestine, he encounters pastoralists (Eg. *styw*) herding cattle, led by a sheikh (*mtn*) and organized into a clan (*whyt*). They themselves had been down into the Delta and so recognize him, and they offer him what we would call typical Bedouin "hospitality." (2) Moving northward to Byblos, then eastward to "Qedum" (*qdm* meaning "east"), Sinuhe arrives at "the land of Yaa/Arau," where he himself becomes a sort of territorial sheikh, under the aegis of his new father-in-law, the Amorite(?) 'Ammu-inshi, ruler (*hqwj*) of "Upper Retenu" or Syria. The description of "figs, vines, olives, fruit trees, wheat and barley, and plentiful cattle" suggests that the "land of Yaa/Arau" was perhaps in the fertile areas of Syria in the Beqà , or the plateau to the east. But references to Sinuhe's "clan" (*whyt*), to raids and skirmishes with nearby tribesmen (*styw*), and to hunting in the desert make it clear that this region bordered on the steppe and was in close contact with pastoral nomads.

The picture derived from the Tale of Sinuhe is thus almost precisely that envisioned by Rowton's "dimorphic society," with its characteristic "enclosed nomadism." And the pastoral morpheme of the story would suit almost perfectly my portrait of the EB IV landscape in Palestine. It is noteworthy that in neither case is there any "economic compulsion" behind the pastoral nomadic lifestyle. Sinuhe deliberately chose it; relished it; persevered in it for many years, despite his obvious love for his Egyptian homeland; and left it reluctantly only in old age. Is there *not* a "cognitive map" here?

Conclusion

Hodder has observed that "the ideational realm is, in most of archaeology, still studied in terms of the functions of symbols and rituals" (1986:154). He remarks, however, that post-processualists "recognize that all archaeologists necessarily impose meaning content, and that such meanings form the core of archaeological analysis which must be made explicit and rigorous." (1986:154). I have tried here to expand the notion of "ideational realm" to the *larger* realm of the "archaeological landscape." While necessarily speculative, I hope that the attempt to utilize the available ethnographic and textual evidence has made this attempt both explicit and rigorous.

Notes

1 Full bibliography on the EB IV period up to about 1990 can be found conveniently in Palumbo 1991, supplemented principally by Dever 1992, 1995; Finkelstein 1992; Cohen 1992; Gophna 1992; Eisenberg 1993; Goren 1996; Haiman 1996.

2 The bibliography on regional archaeology and surface surveys in our branch of archaeology is scant, but see, for example, Dever 1988:342-343.

3 For an orientation to General Systems Theory and an attempt to apply it to Palestinian archaeology (the EB III/IV transition), see Dever 1989. For recent critiques of the basic model, however, see Tilley 1998:311-314. Whitley 1998; 3, 5, 16, 276-278, 311, 312.

4 For orientation see Hodder 1987; Bintliff 1991; Last 1995; Cobb 1998; Peebles 1998 and both references and critique there. For *annales* approaches in our field, cf. Dever 1988.

5 On "collapse" models generally, see Tainter 1989; Yoffee and Cowgill 1988; and for application to Palestinian EB IV cf. Dever 1989.

6 The literature on "post-processualism" is vast, but for orientation see Hodder 1986; 1991; Demarest 1989; Renfrew and Bahn 1991:339-370; Preucel 1991; and Whitley 1998, especially the essays by Flannery and Marcus (1998); Leone (1998), Shanks and Hodder (1998), and Tilley (1998)--all principal spokesmen of the movement. Also see essays in Hodder *et al.* (1995), especially those by Shanks and Tilley; Lucas; Gellner, and Moore. For relatively early approaches to "landscape archaeology," see Tilley 1994; Walsh 1995 (referring to works as early as the 1960's). Cf. also below and n. 17.

7 On the rapid reurbanization of Palestine in MB I (Albright's MB IIA"), see Dever 1987 and references there; and update by reference to Kempinski 1992; Ilan 1995. For an attempt at a quantitative definition of "urbanism," see Falconer 1987, a model that I have adopted in several recent treatments, such as Dever 1997c.

8 The literature on the "Amorites" in the archaeology of Palestine is vast, but for orientation see Dever 1980:53-55; 1992;83, 84 and extensive references there. See also nn. 10, 11 below.

9 The major criticisms have come from my own students, such as Palumbo (1991); Falconer (1987); and Richard (1990); but cf. Goren 1996:67, 68; and also the "multi-morphic" model proposed by Finkelstein (1992). Palumbo and Peterman (1993) have added a "Family AZ," for Amman/'Ain Zerqa in Jordan.

10 On 'Ein Yael, see now Edelstein, Milevski, and Aurant (1998); on Dhâhr Mirzbâneh and other Central Hills sites, mostly cemeteries, see Finkelstein (1991); on the small village at Sha'ar ha-Golan, see Eisenberg (1993). On Tel el-Hayyat, a site where I initiated excavations, as well as nearby Tell Abu en-Niaj, see Falconer and Magness-Gardiner (1989). On Kh. Iskander, see Richard (1990).

11 Palumbo (1991:60), following Richards' early excavations, argues that Kh. Iskander's large city wall was constructed in EB IV. But Richard herself has now shown that the walls in question were built in EB II, and reused only ephemerally, if at all, in EB IV (oral communication). Palumbo's other "walled EB IV towns" (1991:58, 59) also fail to stand up under closer scrutiny.

12 For my responses to these criticisms (cited in nn. 10, 1 above) cf. Dever 1992; 1995. Cf. also n. 9 above, n. 13 below.

13 For the problems of defining "pastoral nomadism," see Johnson 1969; Dever 1977; and now several papers in Bar-Yosef and Khazanov (1992), especially those of Bar-Yosef and Khazanov (1992); Köller-Rollefson (1992); Levy; Marx (1992); and Meadow (1992). Also important are several discussions of "tribes" in Khoury and Kostiner 1990, especially the essays of Khoury and Kostiner (1990); Lapidus (1990); Tapper (1990); Caton (1990); and Gellner (1990). Cf. n 14 below.

14 For the terminology of some of these binary oppositions, see for example Richard 1990:55, 56; LaBianca 1991:38-49; 235-245; Palumbo 1991:127-134; Finkelstein 1992:134; Dever 1995:291, 292. Cf. also n. 13 above.

15 See references to post-processualism in n. 6 above. Most of the essays in Whitley 1998, together with references to other recent literature, convey a clear impression that

"post-processualism" in one form or another now prevails in Americanist and British circles. In Syro-Palestinian archaeology, however, I know of nothing in the literature that even acknowledges the phenomenon, except my own paper (Dever 1993).

16 On "reading material culture as text," see principally Shanks and Tilley 1982; Hodder 1986; 1995a; Tilley 1990; 1991; 1994; Shanks 1992. On the related notion of "writing history from things," see the seminal essays in Lubar and Kingery 1993; Kingery 1996. My own attempt to apply the principle to Palestinian archaeology, and especially to histories of ancient Israel, will be found in Dever 1993; 1997d. On material culture as "primary data: in history-writing, see my treatments in Dever 1997a; 1997b.

17 On "environmental perception," see for example, Tilley 1991:78-86; 1994; Clarkson 1998, especially pp. 119-125. Cf. n. 23 below.

18 Actual *physical* maps of virtually all known EB IV sites will be found in Palumbo 1991, following Pl. 12 (dozens, but unnumbered); cf. also Gophna 1992, p. 129 (but much less complete).

19 For examples of ecological definitions and models, see Johnson 1969:1-3 *et passim*; Bar-Yosef and Khazanov 1992:1-3; Köller-Rollefson 1992; Levy 1992:66; Marx 1992:255; and especially Meadows 1992:262,263.

20 For references, see Dever 1997d:17, 18; Clarkson 1998:122-125; Hosler 1998; Lewis-Williams 1998; and cf. other references in nn. 16,17 above.

21 On the "meaning of things," see references in nn. 16, 17, 20 above. On post-processualism's explicit rejection of the notion that ideology is a "mere epiphenomenon" in cultural change (as in the materialism and environmental determinism of much of the New Archaeology), see Trigger 1989a:23-32; Demarest 1989; Flannery and Marcus 1998:40-43; Leone 1998:50-59.

22 Binford (1989:61) has contemptuously dismissed Hodder and his colleagues as "coggies," because he assumes that "cognitive archaeology" of post-processualism inevitably falls into pure subjectivity and negates the hard-won "scientific" gains of the New Archaeology. For a refutation of this notion, see Lamberg-Karlovsky 1989:6-15; Trigger 1989:23, 24, 344-346; Hodder 1991; Leone 1991; Preucel 1991; McGuire and Saitta 1998:276-278; Shanks and Hodder 1998:87-90; Whitley 1998: 6, 11, 12, 14, 15.

23 See above and n. 17 on "landscape as cultural process." The basic concept of a "cognitive map" may be best expressed by the Arabic term *'assabiya*. The essential meaning is "group solidarity"; but the implied meaning includes the notion of "fanaticism," i.e., fierce devotion to the ideological *basis* of solidarity. In traditional Arabic thought, this concept is best and most purely expressed by the "Bedouin mentality," which is comprised of (1) kinship, based on blood-ties; (2) absolute and unquestioned loyalty to family, tribe, and clan; and autonomy, or resistance to the state's claims to authority, and especially to sedentarization, to becoming peasants or *fellahin*, I would thank my colleague Prof. Adel Gamal for clarification of the Arabic term and concept. He has collected a mass of data on the centuries-old Bedouin mentality"; but he notes that scant literature is available in English. Cf. also nn. 24, 43, 44 below.

24 Cf. n. 23 above. For the prevalence of the notion of resistance to sedentarization--in much of the literature of "tribe-state conflict"--in the literature since Ibn Khaldun, see Barfield 1990:154-164. However, Barfield points out that this view is based on a selective analysis; some "tribal" confederations today are largely sedentary. As many of the essays in Khoury and Kostin (1990) point out, some of these confederations may come to constitute a uniquely Middle Eastern phenomenon, the hybrid "tribal state" (as, for instance, Saudia Arabia until relatively recently). See also Khazanou 1984; Caton 1990:86-90; and cf. n. 25 below.

25 Khoury and Kostiner 1990:19-22, quoting Richard Tapper; Tibi 1990:133, 134.

26 Rowton's seminal articles are scattered throughout several journals, but for references see Rowton 1977.

27 Cf. n. 21 above.

28 Cf. references in n. 8 above.

29 On archaeology and the much-discussed "nomadic ideal" in the Hebrew Bible and in early Israel, see Dever 1997b and references there.

30 For the latest analysis of the Tale of Sinuhe, see Rainey 1998. Rainey agrees in seeing the story as having a real historical context in Syria-Palestine in the early 2nd millennium BCE; but he sees more of an "urban" context than I think is possible.

REFERENCES

Bar-Yosef. O., and Khazanov, A., eds.
1992 *Pastoralism in the Levant: Archaeological Materials in Anthropological Perspectives.* Monographs in World Archaeology 10. Madison: Prehistory Press.

Barfield, T. J.
1990 Tribes and State Relations: The Inner Asian Experience. Pp. 153-182 in *Tribes and State Formation in the Middle East*, eds. P. S. Khoury and J. Kostiner. Berkeley: University of California Press.

Barth, F.
1969 *Nomads of South Persia: The Basseri Tribe of the Khamseh Confederacy.* Boston: Little, Brown and Company.

Binford, L. R.
1972 *An Anthropological Perspective.* New York: Seminar Press.
1989 The "New Archaeology," Then and Now. Pp. 50-62 in *Archaeological Thought in America,* ed. C. C. Lamberg-Karlovsky. Cambridge: Cambridge University Press.

Bintliff, J., ed.
1991 *The Annales School and Archaeology.* London: Leicester University Press.

Caton, S. C.
1980 Anthropological Theories of Tribe and State Formation in the Middle East: Ideology and the Semiotics of Power. Pp. 74-108 in *Tribes and State Formation in the Middle East,* eds. P. S. Khoury and J. Kostiner. Berkeley: University of California Press.

Clarkson, P. B.
1998 Archaeological Imaginings. Pp. 119-130 in *Reader in Archaeological Theory: Post-Processual and Cognitive Approaches,* ed. D. S. Whitley. London: Routledge.

Cobb, C. R.
1998 Social Reproduction and the longue durée in the Prehistory of the Midcontinental United States. Pp. 200-216 in *Reader in Archaeological Theory: Post-Processual and Cognitive Approaches,* ed. D. S. Whitley. London: Routledge

Cohen, R.
1992 The Nomadic or Semi-Nomadic Middle Bronze I Settlements in the Central Negev. Pp. 105-131 in *Pastoralism in the Levant: Archaeological Materials in Anthropological Perspectives,* eds. O. Bar Yosef and A. Khazamov. Monographs in World Archaeology 10. Madison: Prehistory Press.

Demarest, A. A.
1989 Ideology and Evolutionism in American Archaeology: Looking Beyond the Economic Base. Pp. 89-102 in *Archaeological Thought in America,* ed. C. C. Lamberg-Karlovsky. Cambridge: Cambridge University Press.

Dever, W. G.
1971 The Peoples of Palestine in the Middle Bronze I Period. *Harvard Theological Review* 64:197-226.
1973 The EB IV-MB I Horizon in Transjordan and Southern Palestine. *Bulletin of the American Schools of Oriental Research* 210:37-63.
1977 Palestine in the Second Millennium B.C.E.: The Archaeological Picture. Pp. 70-120 in *Israelite and Judean History,* eds. J. H. Hayes and J. M. Miller. Philadelphia: Westminster Press.
1980 New Vistas on the EB IV ("MB I") Horizon in Syria-Palestine. *Bulletin of the American Schools of Oriental Research* 237:35-69.
1987 Palestine in the Middle Bronze Age: The Zenith of the Urban Canaanite Era. *Biblical Archaeologist* 50:149-176.
1988 Impact of the "New Archaeology." Pp. 337-352 in *Benchmarks in Time and Culture: Introduction to Palestinian Archaeology,* eds. J. F. Drinkard, G. L. Mattingly, and J. M. Miller. Atlanta: Scholars Press.
1989 The Collapse of the Urban Early Bronze Age in Palestine: Toward a Systemic Analysis. Pp. 225-246 in *L'urbanisation de la Palestine à âge du Bronze ancien,* ed. P. de Miroschedji. Oxford: BAR International Series.
1992 Pastoralism and the End of the Early Bronze Age in Palestine. Pp. 83-94 in *Pastoralism in the Levant: Archaeological Materials in Anthropological Perspectives,* eds. O. Bar-Yosef and A. Khazanov. Monographs in World Prehistory 10. Madison: Prehistory Press.
1993 Biblical Archaeology--Death and Rebirth? Pp. 706-722 in *Biblical Archaeology Today, 1990. Proceedings of the Second International Congress on Biblical Archaeology, Jerusalem, June 1990,* eds. A. Biran and J. Aviram. Jerusalem: Israel Exploration Society.
1995 Social Structure in the Early Bronze IV Period in Palestine. Pp. 282-296 in *The Archaeology of Society in the Holy Land,* ed. T. E. Levy. Leicester: Leicester University Press.
1997a Philology, Theology, and Archaeology. What Kind of History Do We Want, and What is Possible? Pp. 290-310 in *The Archaeology of Israel: Constructing the Past, Interpreting the Present,* eds. N. A. Silberman and D. Small. Sheffield: Sheffield Academic Press.
1997b Israelite Origins and the "Nomadic Ideal": Can Archaeology Separate Fact from Fiction? Pp. 220-237 in *Mediterranean Peoples in Transition: Thirteenth to Tenth Centuries B.C.E.,* eds. S. Gitin, A. Mazar, and E. Stern. Jerusalem: Israel Exploration Society.
1997c Archaeology and "Age of Solomon": A Case-Study in Archaeology and Historiography. Pp. 217-251 in *The Age of Solomon: Scholarship of the Turn of the Millennium,* ed. L. K. Handy. Leiden: E. J. Brill.
1997d On Listening to the Text--And the Artifacts. Pp. 1-23 in *Echoes of Many Texts: Reflections on Jewish and Christian Traditions,* eds. W. G. Dever and J. E. Wright. Atlanta: Scholars Press.

Edelstein, G., Milevski, I, and Aurnat, S.
1998 *Villages, Terraces and Stone Mounds. Excavations at Manahat, Jerusalem*. 1987-1989. Jerusalem: Israel Antiquities Authority.

Eisenberg. E.
1993 Sha'ar ha-Golan. Pp. 1340-1343 in *The New Encyclopedia of Archaeological Excavations in the Holy Land*, ed. E. Stern. Jerusalem: Israel Exploration Society.

Falconer, S.
1987 *Heartland of Villages: Reconsidering Early Urbanism in the Southern Levant*. Unpublished doctoral dissertation, University of Arizona.

Finkelstein, I.
1991 The Central Hillcountry in the Intermediate Bronze Age. *Israel Exploration Journal* 41:14-45.
1992 Pastoralism in the Highlands of Canaan in the Third and Second Millennia B.C.E. Pp. 133-142 in *Pastoralism in the Levant: Archaeological Materials in Anthropological Perspectives*, eds. O. Bar-Yosef and A. Khazanov. Monographs in World Archaeology 10. Madison: Prehistory Press.

Flannery, K. V., and Marcus. J.
1998 Cognitive Archaeology, Pp. 35-48 in *Reader in Archaeological Theory: Post-Processual and Cognitive Archaeology*, ed. D. S. Whitley. London: Routledge.

Gellner, E.
1990 Tribalism and the State in the Middle East. Pp. 109-126 in *Tribes and State Formation in the Middle East*, eds. P. S. Khoury and J. Kostiner. Berkeley: University of California Press.

Gophna, R.
1992 The Intermediate Bronze Age. Pp. 126-158 in *The Archaeology of Ancient Israel*, ed. A. Ben-Tor. New Haven: Yale University Press.

Goren, Y.
1996 The Southern Levant in the Early Bronze Age IV: The Petrographic Perspective. *Bulletin of the American Schools of Oriental Research* 303:33-72.

Haiman, M.
1996 Early Bronze Age IV Settlement Pattern of the Negev and Sinai Deserts: View from Small Marginal Temporary Sites. *Bulletin of the American Schools of Oriental Research* 303:1-32.

Hodder, I.
1986 *Reading the Past: Current Approaches to Interpretation in Archaeology*. Cambridge: Cambridge University Press.
1987 Converging Traditions: the Search for Symbolic meanings in Archaeology and Geography. Pp. 134-145 in *Landscape and Culture: Geographical and Archaeological Perspectives*, ed. J. M. Wagstaff. Oxford: Basil Blackwell
1991 Postprocessual Archaeology and the Current Debate. Pp. 30-41 in *Processual and Postprocessual Archaeologies: Multiple Ways of Knowing the Past. Center for Archaeological Investigations, Occasional Paper 10.* Carbondale: Southern Illinois University.

Hodder, I., ed.
1987 *Archaeology as Long-Term History*. Cambridge: Cambridge University Press.
1995 *Theory and Practice in Archaeology: Material Culture*. London: Routledge.

Hodder, I. et al.
1995 *Interpreting Archaeology: Finding Meaning in the Past*. London: Routledge.

Hosler, D.
1998 Sound, Color and Meaning in the Metallurgy of Ancient West Mexico. Pp. 103-118 in *Reader in Archaeological Theory: Post-Processual and Cognitive Approaches*, ed. D. S. Whitley. London: Routledge.

Ilan, D.
1995 The Dawn of Internationalism--The Middle Bronze Age. Pp. 197-319 in *The Archaeology of Society in the Holy Land*, ed. T. E. Levy. Leicester: Leicester University Press.

Johnson, D. L
1969 *The Nature of Nomadism: A Comparative Study of Pastoral Migrations in Southwestern Asia and Northern Africa*. University of Chicago, Department of Geography Research Paper 118. Chicago: University of Chicago.

Kempinski, A.
1992 The Middle Bronze Age. Pp. 159-210 in *The Archaeology of Ancient Israel*, ed. A. Ben-Tor. New Haven: Yale University Press.

Khazanov, A.
1984 *Nomads and the Outside World*. Cambridge: Cambridge University Press.

Khoury, P. S., and Kostiner, J., eds.
1990 *Tribes and State Formation in the Middle East*. Berkeley: University of California Press.

Kochavi, M.
1967 *The Settlement of the Negev in the Middle Bronze (Canaanite) I Age*. Unpublished doctoral dissertation, The Hebrew University of Jerusalem.

Köhler-Rollefson, I.
1992 A Model for the Development of Nomadic Pastoralism on the Transjordan Plateau. Pp. 11-18 in *Pastoralism in the Levant: Archaeological Materials in Anthropological Perspectives*, eds. O. Bar-Yosef and A. Khazanov. Monographs in World Archaeology 10 Madison: Prehistory Press.

Kingery. W. D.
1996 *Learning From Things: Method and Theory of Material Culture Studies.* Washington: Smithsonian Institution.
LaBianca, S.
1990 *Hesban I. Sedentarization and Nomadization: Food System Cycles at Hesban and Vicinity in Transjordan.* Berrien Springs: Andrews University.
Lapidus, I. M.
1990 Tribes and State Formation in Islamic History. Pp. 25-47 in *Tribes and State Formation in the Middle East*, eds. P. S. Khoury, and J. Kostiner. Berkeley: University of California Press.
Lapp, P. W.
1970 Palestine in the Early Bronze Age. Pp. 101-131 in *Near Eastern Archaeology in the Twentieth Century*, ed. J. A. Sanders. Garden City: Doubleday.
Lamberg-Karlovsky, C. C., ed.
1989 *Archaeological Thought in America.* Cambridge: Cambridge University Press.
Last. J.
1995 The Nature of History. Pp. 141-157 in *Finding Meaning in the Past*, ed. I. Hodder et al. London: Routledge
Leone, M. P.
1991 Materialist Theory and the Formation of Questions in Archaeology. Pp. 235-241 in *Processual and Postprocessual Archaeologies: Multiple Ways of Knowing the Past*, ed. R. W. Preucel. Center for the Archaeological Investigations, Occasional Paper 10. Carbondale: University of Southern Illinois
1998 Symbolic, Structural, and Critical Archaeology. Pp. 49-68 in *Reader in Archaeological Theory: Post-Processual and Processual Approaches*, ed. D. S. Whitley. London: Routledge.
Levy, T. E.
1991 Transhumance, Subsistence, and Social Evolution in the Northern Negev Desert. Pp. 65-82 in *Pastoralism in the Levant: Archaeological Materials in Anthropological Perspectives*, eds. Bar- Yosef and A. Khazanov. Monographs in World Archaeology 10. Madison: Prehistory Press.
Lewis-Williams, J. D.
1998 Wrestling With Analogy. Pp. 157-175 in *Reader in Archaeological Theory: Post-Processual and Cognitive Archaeologies*, ed. D. S. Whitley. London: Routledge.
Lubar, S., and Kingery, W. D., eds.
1993 *History from Things: Essays on Material Culture.* Washington: Smithsonian Institution.
Matthews. V. H.
1978 *Pastoral Nomadism in the Mari Kingdom (ca. 1830-1760 B.C.).* Cambridge: American Schools of Oriental Research.
Marx. E.
1992 Are There Pastoral Nomads in the Middle East? Pp. 255-260 in *Pastoralism in the Levant: Archaeological Materials in Anthropological Perspectives*, eds. O. Bar-Yosef and A. Khazanov. Monographs in World Archaeology 10. Madison: Prehistory Press.
Meadow, R. H.
1992 Inconclusive Remarks on Pastoralism, Nomadism, and Other Related Matters. Pp. 261-269 in *Pastoralism in the Levant: Archaeological Materials in Anthropological Perspectives*, eds. O. Bar-Yosef and A. Khazanov. Monographs in World Archaeology 10. Madison: Prehistory Press.
Palumbo, G.
1991 *The Early Bronze Age IV in the Southern Levant: Settlement Patterns, Economy, and Material Culture of a "Dark Age." Contributi e Materials di Archeologia Orientale III.* Rome: University of Rome.
Palumbo, G., and Peterman, G.
1993 Early Bronze Age IV Regionalism in Central Jordan. *Bulletin of the American Schools of Oriental Research* 289:23-32.
Peebles, C. S.
1998 Annalistes, Hermeneutics and Positivists: Squaring circles or Dissolving Problems. Pp. 183-197 in *Reader in Archaeological Theory: Post-Processual and Cognitive Approaches*, ed. D. S. Whitley. London: Routledge.
Preucel, R. W., ed.
1991 *Processual and Postprocessual Archaeologies: Multiple Ways of Knowing the Past. Center for Archaeological Investigations, Occasional Paper 10.* Carbondale: Southern Illinois University.
Rainey, A. F.
1998 "This is the Taste of Death": A Fleeing Egyptian Bureaucrat Reveals What Life in Ancient Canaan Was Like. *Archaeological Odyssey* 1/4:42-47; 68, 69.
Renfrew, C., and Bahn, P.
1991 *Archaeology: Theories, Methods, and Practice.* New York: Thames and Hudson.
Richard, S.
1990 The 1987 Expedition to Khirbet Iskander and Its Vicinity: Fourth Preliminary Report. Pp. 33-58 in *Preliminary Reports of ASOR-Sponsored Excavations, 1983-87*, ed. W. E. Rast. Baltimore: American Schools of Oriental Research.

Rowton, M. S.
1977 Dimorphic Structure and the Parasocial Element. *Journal of Near Eastern Studies* 36:181-198.

Salzman, P. C., ed.
1980 *When Nomads Settle: Processes of Sedentarization as Adaptation and Response*. New York: Praeger.

Shanks, M.
1992 *Experiencing the Past: On the Character of Archaeology*. London: Routledge.

Shanks, M., and Hodder, I.
1998 Processual, Postprocessual and Interpretive Archaeologies. Pp. 69-95 in *Reader in Archaeological Theory: Post-Processual and Cognitive Archaeologies*, ed. D. S. Whitley. London: Routledge.

Shanks, M., and Tiley, C.
1987 *Re-Constructing Archaeology: Theory and Practice*. Cambridge: Cambridge University Press.

Tainter, J. A.
1988 *The Collapse of Complex Societies*. Cambridge: Cambridge University Press.

Tibi, B.
1990 The Simultaneity of the Unsimultaneous: Old Tribes and Imposed Nation-States in the Modern Middle East. Pp. 127-152 in *Tribes and State Formation in the Middle East*, eds. P. S. Khoury and J. Kostiner. Berkeley: University of California Press.

Tilley, C.
1990 *Reading Material Culture: Structuralism, Hermenéutics and Post-Structuralism*. Oxford: Basil Blackwell.
1991 *Material Culture and Text: The Art of Ambiguity*. London: Routledge.
1994 *A Phenomenology of Landscape: Places, Paths and Monuments*. Oxford: Berg.
1998 Archaeology as Socio-political Action in the Present. Pp. 315-330 in *Reader in Archaeological Theory: Post-Processual and Cognitive Approaches*, ed. D. S. Whitley. London: Routledge.

Tapper, R.
1990 Anthropologists, Historians, and Tribes people on Tribe and State Formation in the Middle East. Pp. 48-73 in *Tribes and State Formation in the Middle East*, eds. P. S. Khoury and J. Kostiner. Berkeley: University of California Press.

Trigger, B. D.
1989a History and Contemporary American Archaeology: A Critical Analysis. Pp. 19-34 in *Archaeological Thought in America*, ed. C. C. Lamberg-Karlovsky. Cambridge: Cambridge University Press.
1989b *A History of Archaeological Thought*. Cambridge: Cambridge University Press.

Walsh, K.
1995 A Sense of Place: A Role for Cognitive Mapping in the "Post-modern" World? Pp. 131-140 in *Finding Meaning in the Past*, eds. I. Hodder et al. London: Routledge.

Whitley, D. S.
1998 *Reader in Archaeological Theory: Post-Processual and Cognitive Approaches*. London: Routledge.

Yoffee, N.; and Cowgill, G. L., eds.
1988 *The Collapse of Ancient States and Civilizations*. Tucson: University of Arizona Press.

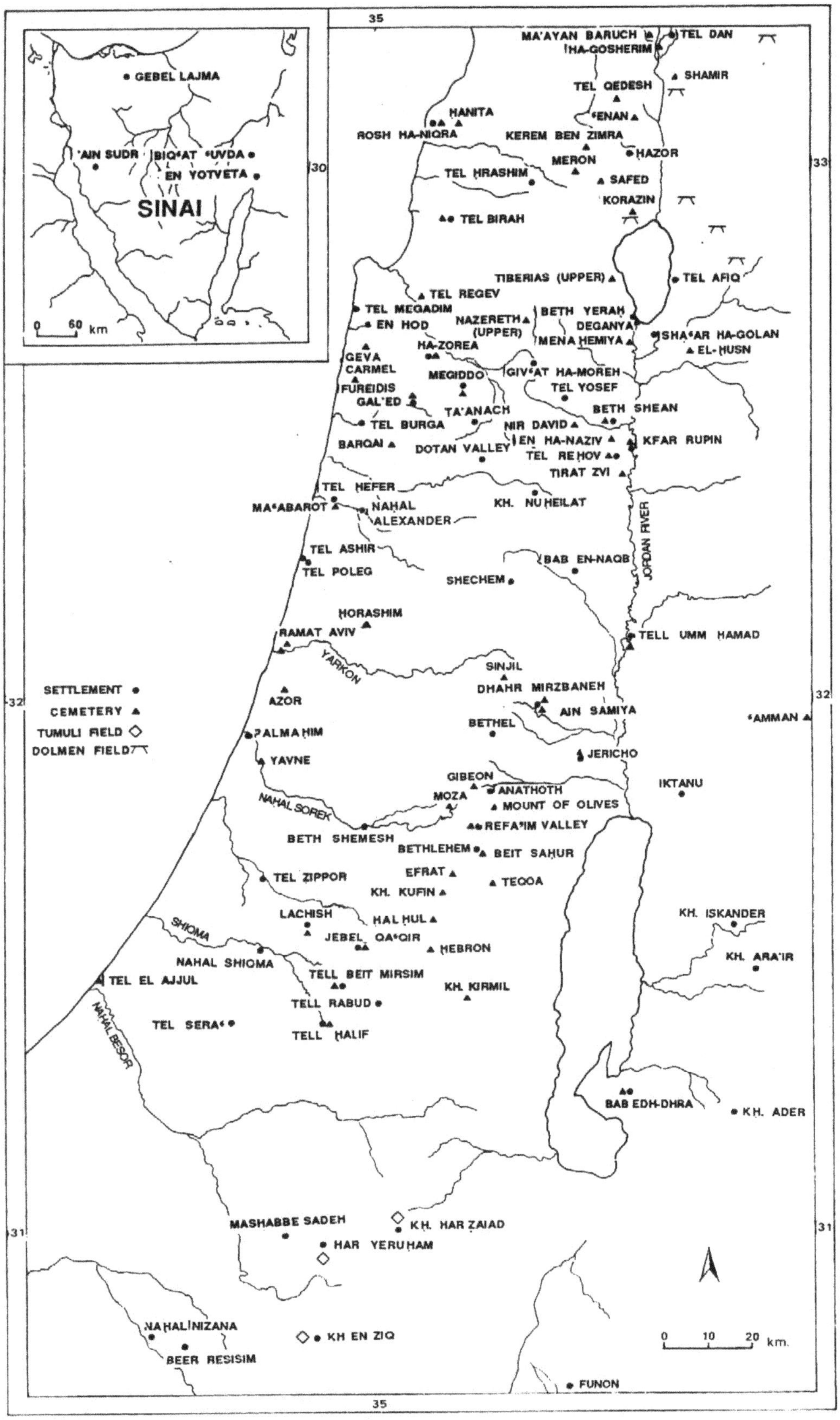

Figure 3-1: Early Bronze IV sites and cemeteries (Gophna 1992, Figure 5.1)

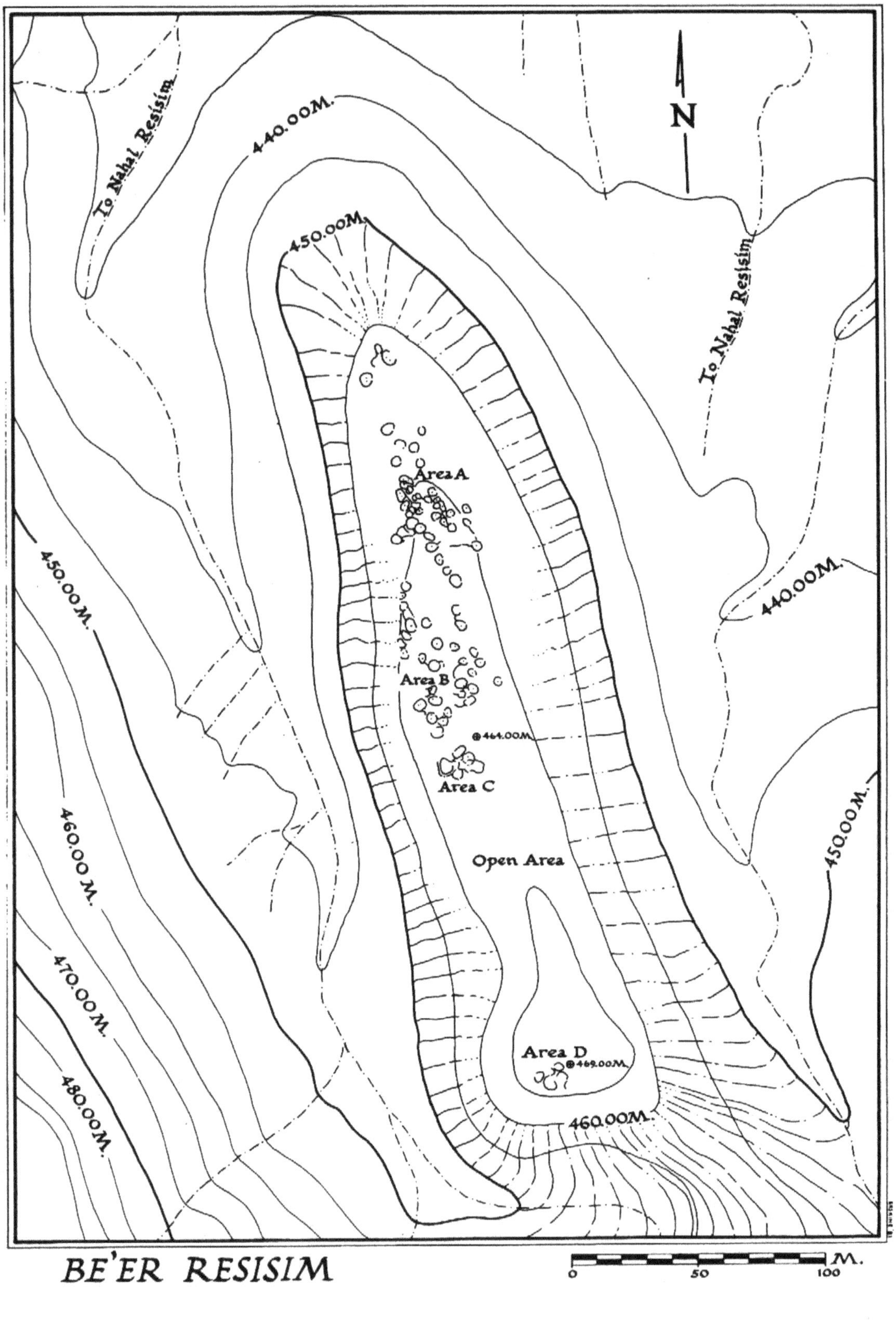

Figure 3-2a: Contour map of the ridge and village plan of the main settlement, showing all excavated areas (plan by Israel Vatkin and Robert Erskine)

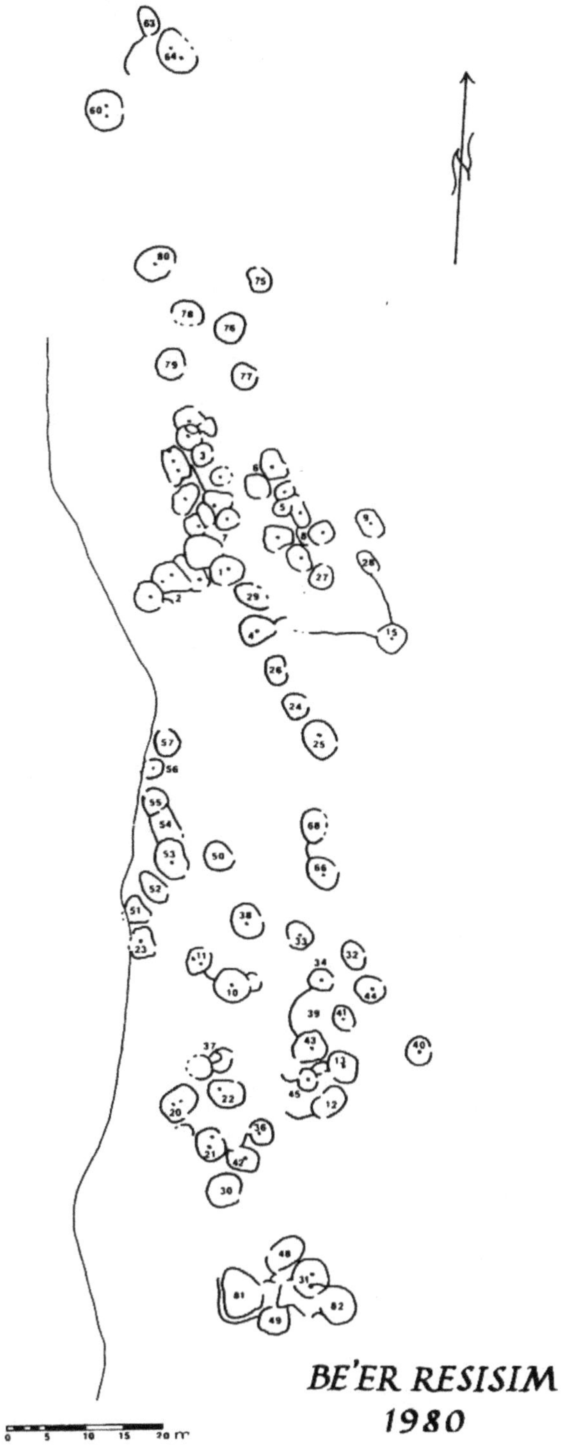

Figure 3-2b: Enlarged plan of the structures of Areas A-C at the completion of excavation (plan by Israel Vatkin)

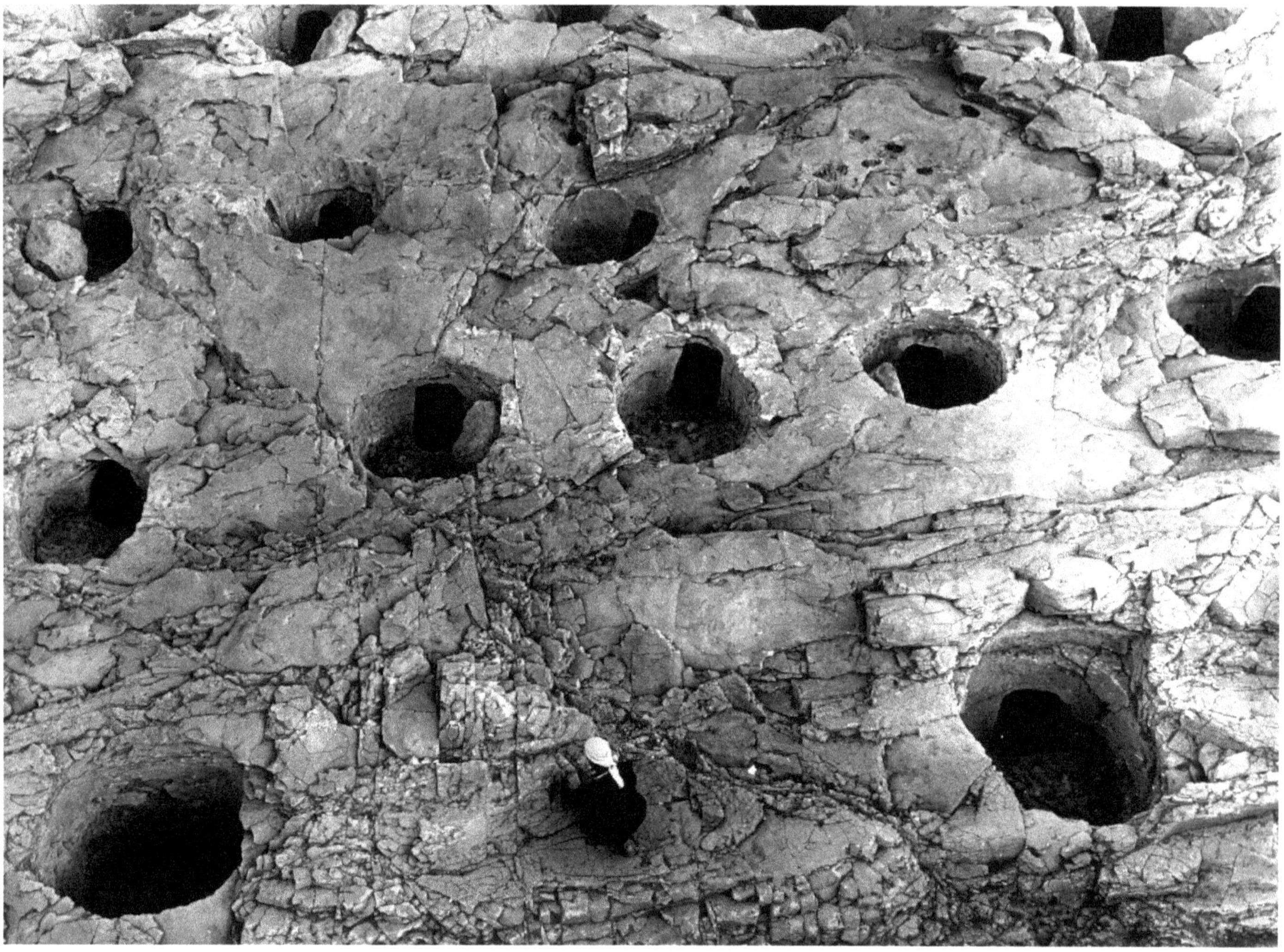

Figure 3-3: Jebel Qa'aqir: EB IV shaft-tombs in Cemetery B, 2000 B.C. (photo by Theodore Rosen).

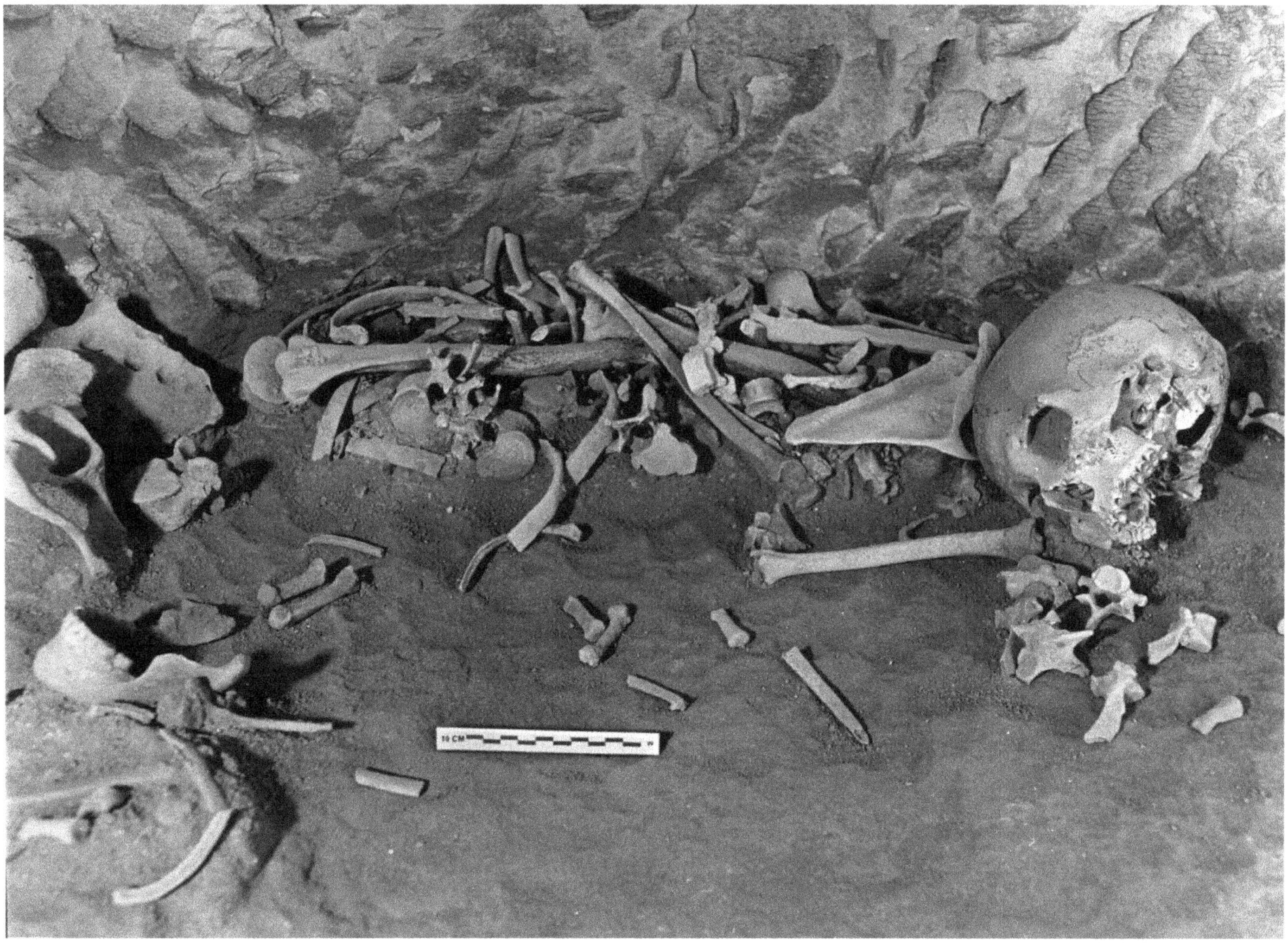

Figure 3-4: Disarticulated secondary burial in Tomb B47 at Jebel Qa'aqir (photo by Theodore Rosen)

Figure 3-5: Pastoral nomads on the move near Mari on the Euphrates (photo by W.G. Dever)

Does Size Count? Urban and Cultic Perspectives on the Rural Landscape during the Middle Bronze Age II[1]

AREN M. MAEIR

The Institute of Archaeology
The Martin (Szusz) Department
of the Land of Israel Studies
and Archaeology
Bar Ilan University
Ramat Gan, 52900
Israel
maeira@mail.biu.ac.il

In recent years, several studies have appeared that have attempted to downplay the role of the urban elements (and thus, overemphasize the role of the rural components) of the Middle Bronze Age II settlement in the southern Levant. It has even been claimed that there were no urban sites during this period. In the following pages, this claim is contended, and it is shown that urban sites fullfilled a central and undeniable role within the Middle Bronze Age II settlement matrix. This point is argued through a) a discussion of the relative size and role of the various sites; and b) a review of the "sacred landscape" during this period.

Introduction

This paper is somewhat of an anomaly for this volume. As one would expect from the title of this collection, the majority of the papers deal directly with the rural components of the land, covering a wide range of periods. In the following, I have chosen to discuss aspects relating to rural settlement in the southern Levant, but from a different angle. I will examine the rural component in the Land of Israel during the Middle Bronze Age II (MB II), not with an emphasis on the rural sites themselves, but rather by a) observing their relationships to the urban entities during this period; and b) exploring the role and relationship of rural and urban cultic centers of the MB II.

As was pointed out both in the introduction to the volume and in several of the studies, it is only in the more recent archaeological research in the southern Levant has sufficient attention been paid to the smaller, primarily rural sites. Earlier research had for the most part focused on the large sites, especially tels (e.g. Ahlström's "tell-minded archaeology" [1982:35]). Over the last several decades, we have witnessed an increased awareness of the need to study the smaller sites as well. In addition, much emphasis has recently been placed on regional studies, in which the various classes, types and sizes of sites are studied in an intergrated, holistic manner.

This being said, it would be unwarranted to over-emphasize the rural component and underestimate the importance of the urban setting. Although one cannot truly understand the ancient cultures without the study of the rural component, the urban component is of no less importance.

Several studies in the last decade or so have emphasized the importance of the rural component during the MB II. In fact, it has been suggested in some of these studies, that the rural component is to be understood as the permanent constituent in the settlement pattern of both this and other periods, as opposed to the "transient" urban settlements (e.g., Palumbo 1990; Joffe 1991; 1993; for a critical discussion of these opinions, see Greenberg, this volume). Similarly, London (1992) argues that the large tels, usually interpreted as full-fledged urban entities, served solely as sites at which the ruling elite and their immediate subordinates resided. Accordingly, the majority of the population lived in the smaller, mainly rural sites, where most of the economic activities were conducted.

It is precisely this perspective that I would like to contend. I have no intention of claiming that the rural component of ancient settlement patterns is not important. Rather, I strongly believe that the above-mentioned (and other) recent studies that have accentuated the rural aspect (often due to current fads in archaeological interpretation) have, in fact, relegated the urban component to a subsidiary position, which is unwarranted.[2] This point is argued based on data relating to the Jordan Valley during the MB II.

The Urban-Rural Matrix in the Central Jordan Valley

A striking example, I believe, of the over-emphasis on the rural component, can be found in Falconer's studies of the Jordan Valley during the MB II (1987a;1987b), in which he posits that all the MB II sites in the Jordan valley (and for that matter in most of the Land of Israel) were villages, not cities.[3] According to this view, all the site-functions that one encounted in the archaeological record were carried out in various manifestations of the village-type settlement.

Based on the results and interpretations of an intensive and extensive regional study of the central Jordan Valley, in which particular emphasis was placed on the remains dating to the MB II (Maeir 1997a) , I believe that this view is seriously flawed. It is based on an essential misconception - the question of the comparative size of the respective sites. Falconer (1987a; 1987b) compares the sizes of the MBII sites in the Jordan Valley in particular, and in the southern Levant in general, to the sizes of several ancient Mesopotamian cities.[4] Due to the relatively minuscule size of the southern Levantinian sites, he suggests that they should not be understood as urban structures, but rather as village-based entities.

Such a comparison does not seem viable. It is mistaken to assume one can determine a site's function on the basis of a comparison between the sizes of different settlement entities in disparate cultures and settlement constellations. If one accept's Falconer's opinion, only one city (Hazor) would have existed in the entire Land of Israel during the MB II (on the political and economic role of Hazor, see Maeir 2000a).

The implausibilty of comparing the sizes of ancient sites in the southern Levant with those of Mesopotamia was already been noted by Dever (1993). Contrary to Falconer, Dever intuitively suggests that the approximate size of an urban unit in the MB II Levant is 6-8 hectare (1993:99-100). Although he does not demonstrate the validity of this suggestion, it does seem to withstand scrutiny (for a discussion, see Maeir 1997a:230-232; 1998). On this basis, one can reconstruct a sufficient number of urban centers in the various regions of the country, each fulfilling a distinct regional and supra-regional role.

In examining the settlement pattern in the southern Levant during the MB II, one must assume the existence of urban centers alongside the rural entities. Mabry (1985), Bunimovitz (1989; 1992a), and this author (Maeir 1997a; 1998) have discussed in detail the multi-scaled components of this framework. It is absolutely clear that there were multi-tiered settlement patterns on the sub-, intra-, and inter-regional levels, and there is no way in which one can comprehend these hierarchal structures unless one assumes varying levels of functional diversity among sites of different sizes. As in any classic settlement pattern (e.g., Christaller 1966; Hodder and Orton 1976; C.A. Smith 1976), this hierarchy includes upper-echelon sites (of varying sizes) that played central, regional role, alongside smaller sites that fullfiled any of a variety of subsidiary roles. Although the smaller sites did have functions beyond supplying their immediate subsistence needs (e.g., pottery production for regionally-oriented consumption [Falconer 1987b]), they were subordinate to the larger sites from a political and economic point of view (on the economic structure of the MB II southern Levant, see Maeir 2000a; Holladay 2001).

It has been repeatedly demonstrated in numerous instances that within a given settlement framework, sites with *similar* functions may be of *varying* sizes (for comments on this problem, see, e.g., Renfrew and Level 1979). When dealing with settlement patterns and types in the Levant (especially regarding the southern Levant), one must take into consideration the specific regional framework. Despite their supposedly minuscule sizes, the urban centers of this region fullfilled roles in relation to their surroundings quite similar to those fulfilled by the much larger urban centers of the Syro-Mesopotamian world. It is reasonable to assume that if a site performs urban functions, it should be regarded as an urban entity, regardless of its size (e.g., Blanton 1976:250-253; M.S. Smith 1989). In the geographical literature, it has been repeatedly demonstrated that the functional differentiation of settlements is a remarkably complex issue, even within the same cultural framework (e.g., Moser and Scott 1961), not to mention when different cultures are compared. Clearly, a somewhat vague comparison between Syro-Mesopotamian and Levantine urban patterns cannot withstand close scrutiny.

As has been pointed out in the past, settlement size is not the only criteria that can, and should, be used to define a settlement as being urban in nature. In fact, it can be argued that overall size is not valid criterion in many cases. The best criterion by which to define a city is most probably functionality. Thus, a settlement should be defined as a city on the basis of the specialized functions it fulfills in relation to its surroundings, and the extent of social stratification and specialization that is observed within (for recent discussions, see, e.g. McIntosh and McIntosh 1993:625; Knapp 1997:56). Therefore, to reiterate an attempt to define whether there were urban entities within the social and economic structure of the MB II southern Levant cannot be based solely on a comparison with site-sizes in other urban cultures. Furthermore, if one opted for Falconer's (1987a) approach, according to which the numerous sites are, for the most part, to be seen as self-sufficient entities, the entire inter-site relationship system among the MB II sites in the southern Levant (hinted to, for example, from historical sources such as the Mari letters) would be untenable. This claim is also in complete contradiction to later historical sources from the Late Bronze and Iron Ages.[5]

A case in point could be the comparison between the urban centers of the Middle Kingdom/Second Intermediate Period (contemporaneous with the MB II) in Egypt and Nubia. As O'Connor demonstrates, while the major (and especially royal) urban sites in Egypt were extremely large (100-225 hectares), the unquestionably urban sites in Kerma (and for that matter some of the smaller Egyptian urban sites as well) were ca. 6-13 hectares in size (O'Connor 1991:156). This seems to be comparable to the situation in the southern Levant discussed above.

Zaccagnini (1989:53) claims that village-based (as opposed to urban) settlement frameworks flourished in areas that could not support intense, irrigated cereal cultivation. I believe that this supposition hardly applies to the southern Levant, particularly in the later, better documented periods, where there is ample evidence of urban structures without the existence of irrigation agriculture. In any event, there most probably was some sort of irrigation agriculture in the Jordan Valley during the MB II (Maeir 1997a),which only strengthens the argument that an urban-centered socio-economic system existed in the region.

Despite these reservations, Falconer's (1987a) thesis regarding the centrality of the rural economic framework in the southern Levant, can be useful. As shown, for example, by Wells (1984), the economic functions of larger settlements are often fulfilled in smaller settlements as well. Thus, while it appears foolhardy to discard the role of urban structures in the ancient Levant, there is no doubt that the central role of rural settlements must be recognized. In this view, the rural sites in the region were part of a prevalent urban/rural framework. It would seem that rather than stress the predominance of one mode of existence at the expense of the other, the urban/rural matrix should be seen as complementary parts of one system. At different times, depending on the socio-economic situation, the respective components would be amplified and/or reduced accordingly.[6] An example of this phenomenon can be seen in Graffam's (1992) discussion of such a relationship under conditions of societal collapse. He notes that when a state-based society collapses, it is never sweeping; rather, the rural components of the society are intensified to offset the negative trends. Recently, Dessel (1999) has argued for a similar interpretation regarding the role and function of rural sites during the transition between the Late Bronze and Iron I periods in the Land of Israel. He stresses that various functions traditionally seen as urban-based also existed in the rural sphere.

"Sacred Landscape" and the Urban/Rural Matrix

An additional facet of urban/rural dynamics can be observed through the study of "Sacred Landscape" (Alcock 1993). This will be approached by focusing once again on the regional distribution of cultic remains in the central Jordan Valley (for a general survey, see Maeir 1997a). This requires attempting to define the placement and roles of the various cultic sites in the region, and to incorporate them into the overall cultural context of the Jordan Valley in the MB II. Alcock's (1993:142-172) definitions of "centralized" and "rural" cults are useful. Centralized cults are cults within urban centers, while the rural cults constitute a wide variety of cultic sites in non-urban settings.

As surveyed previously (Maeir 1997a), cultic remains dating to the MB II were found at the following sites in the Central Jordan Valley: Pella, where these is evidence of a large-scale cultic center apparently founded in the late MB II (S.J. Bourke, oral communication); Tel Beth-Shean, which yielded assorted evidence of cultic activities (Maeir 1997a:76); Tell el-Hayyat, with a temple that was in use throughout the entire MB II (Magness-Gardiner and Falconer 1994); Tel Kitan, with a late MB II temple (Eisenberg 1993a; Maeir 1997a:218-224); and the "Kfar Ruppin site", which has apparent evidence of a temple (Gophna 1979). In other words, these cultic foci are situated at different types of sites: large, central sites (Pella and assumedly at Rehov as well [Mazar 1999]), mid-size sites (Beth-Shean), and smaller, rural sites (Hayyat, Kitan, and the "Kfar Ruppin site"). Cultic centers at the large and medium-size MB II sites throughout the southern Levant are well-known (e.g., Mazar 1991; Katz 2000) and are not be discussed here . Similarly, several rural cultic sites (both temples and less-defined cultic foci) have been reported from sites outside the Jordan Valley.[7]

As mentioned above, the ritual matrix in the central Jordan Valley is comprised of both urban and rural cults, and it conforms nicely with Alcock's (1993) attempts to construct a "Sacred Landscape." In her study, she demonstrates that there is a constant and very complex interplay between urban and rural ritual elements. These relationships can serve as excellent gauges of stability and change, political and economic influence and other factors. It is precisely these directions of inquiry that are relevant in the present study.

Starting from quite a similar premise, Magness-Gardiner and Falconer's (1994) recently discussed the role of the temple at Tell el-Hayyat within the central Jordan Valley regional rural/urban matrix. It appears, however, that two of the basic premises of their analysis are debatable. First, they over-emphasize the importance of the site. I see no reason why Tell el-Hayyat should be interpreted as playing a central role within the central Jordan Valley. The site is one of many small- to medium-sized sites in the region (Maeir 1997a; 2000b). The only outstanding aspect is that it was excavated. Such small-scale ritual buildings in small-size settlements are a common phenomenon in the southern Levant in ancient as well as modern times. The existence of such a temple cannot be used as barometer for site-centrality or importance. As in the modern-day Middle East, where even the smallest of settlements most often has a mosque (Baer 1982:9; Grossman 1994:41), temples in rural settings in the Bronze Age Levant should be seen in a similar light.

Second, Magness-Gardiner and Falconer (1994) claim that the finds from the Tell el-Hayyat temple reflect the social, economic and ritual needs of a rather autonomous village community. Although buttressed by an impressive array of data, I believe that this supposition has not been proven. In fact, this interpretation seems to be an offshoot of Falconer's over-emphasis of the rural component that was questioned above. Conversely, it can be argued that in fact, the various evidence from the temple may indicate external control/influence at the site.

As often noted in the past (e.g., Knapp 1988; Trigger 1990; Pollock 1999: 173-195; Rappaport 1999:313-343), cultic/ideological frameworks are frequently utilized to legitimize the power structures of political authorities. I would suggest that a similar situation existed in the central Jordan Valley, and as a result of this, the urban- and rural-based cults are in fact closely interrelated. The quite similar cultic material culture found at urban and rural sites is telling, as evidenced for example by the similar architecture and cultic paraphernalia both at sites in the central Jordan Valley and at other MB II sites. These overt similarities in cultic material remains at both small and large sites permits one, I believe, to assume that there were strong ritual/ideological relationships among the various classes of sites (e.g., Renfrew 1994:53-54). It should be stated that in fact there are no written documents from the MB II Southern Levant mentioning direct connections and dependance between the rural and urban sites vis-a-vis the cultic facets. Nevertheless, based on the many parallels both from roughly contemporary cultures in the region (e.g. during the Iron Age II, see Miller [2000: 76-79] for a discussion of the local and regional cults during this period), from disparate cultures (e.g. Hindu temples, see Fuller 1988:63), as well as from cultures "more close to home" (e.g. modern Western religions), it is highly likely that the local, largely rural cultic sites were intimately connected, and to a large extent subservient, to their very similar, albeit larger urban counterparts. Such a connection would then be a reflection of political domination, the inter-related cultic/ideological framework serving as a legitimizing force of such dominance.[8] Needless to say, this is intimately connected with the multi-tiered socio-economic, political and settlement structure throughout the southern Levant in general and the central Jordan Valley in particular during the MB II (e.g., Ilan 1995a; Maeir 1997a; 1997b; 2000a). The highly hierarchal, non-egalitarian structure of MB II society is evidenced, for example, by the impressive "public works" (fortifications, temples, palaces) however their function is explained (e.g., Dever 1987; Bunimovitz 1992b; Finkelstein 1992). In addition, the substantial rise in the choice and availability of prestige items, as well as what appears to be differential access to these symbols of prestige is indicative of these trends (e.g. Maeir 1997b). A case in point demonstrating the limited access to prestige items are the metal weapons of the period. Although there is archaeological and documentary evidence of the widespread use of and international trade in bronze weapons and tools (e.g., Dossin 1970; Maeir 2000a), relatively few examples have actually been found in the MB II strata and tombs in the southern Levant. Philip (1991:89) and Ilan (1995a:313) have argued that this should be seen as evidence of elite regulation and control of these resources.[9]

In light of the above, it can be assumed with some degree of certainty that during the MB II, ideological/cultic frameworks existed that served primarily to promote and legitimize elite dominance in MB II society.[10] Furthermore, it is to be expected that the cultic matrix/sacred landscape would mirror the socio-economic and political structure. Thus, the large centers in the central Jordan Valley (Pella and Rehov [Maeir 1997a]) would have larger centralized cults, while the smaller sites, which were dominated by these larger centers, would have cultic foci closely associated with the cults practiced at the central sites.[11] If this suggestion is accepted, then it can be assumed that the Tell el-Hayyat temple was most probably part of the Pella polity, while the Beth-Shean, Tel Kittan and "Kfar Ruppin" temples were most probably incorporated into the Rehov polity. In fact, it would be more than reasonable to assume that the ideological foundations of these respectively related cults would be polity-oriented. Interestingly, the three rural cultic sites in the central Jordan Valley seem to be situated just on the border between the two polities of Pella and Rehov, and all three are in the immediate vicinity of the Jordan River, which most probably served as the border between these two entities. If, in fact, these rural cultic sites served to demarcate this border, these small, rural sites had an additional function (besides serving as cultic centers for these sites, which based on the definition above substantiates the urban-rural heirarchy), serving as so-called "border sanctuaries" (Mazar 1990:496; Alcock 1993:208; Knapp 1999:248. Cf. Haran 1978:56, n. 22). Thus, they would have played an additional role in substantiating the political and ideological framework of the interests of the dominant urban polities. Needless to say, this would only strengthen the claim that they are but part of a "sacred landscape" that was dominated by "elite," urban entities.

Summary

The purpose of this study is not to deny the importance and significance of the rural components of society during the MB II. Rather, as a reaction to the arguments presented in several previous studies, I proposed that one should not over-emphasize the rural components, particularly if this involves leaving the urban component at the wayside. One cannot assume that the urban component was either non-existent, or, if at all, insignificant. It is hardy acceptable that geographical and economic models of urban-rural relations that have been demonstrated in a wide range of cultural, spatial and temporal contexts should be totally irrelevant for the study of MB II society. Rather, one should see these two components as highly integrated, symbiotic facets of the MB II cultural, social and economic fabric.

Needless to say, one must keep in mind the different scales with which one is dealing when comparing geographic matrices in different parts of the ancient Near East. Thus, while the average size of urban units in Mesopotamia is much larger than that in the southern Levant, this does not mean that only sites in the southern Levant that are as large as the Mesopotamian cities should be considered urban in nature. Rather, one should examine the geographic matrix of the MB II settlements in the southern Levant from a more holistic viewpoint.[12] The varied sizes, types and functions of these settlements necessitate that they be interpreted within the context of an urban-rural relationship. Similarly, the varied types and sizes of cultic foci do not suggest the presence of numerous autonomous cultic/ideological centers, but rather to a web of closely interrelated and hierarchical structured entities. Just as the control of land and other economic aspects were firmly entrenched in the hierarchical social and economic structure (e.g., Magness-Gardiner 1994), as part and parcel of the urban/rural relationship, there is no reason to assume that the cultic sphere was any different.
Returning to the question in the title -- Yes! Size does count, but only from a relative and contextual perspective!

Notes

1 This paper is an updated and substantially revised version of portions of Chapter 6 of my Ph.D. dissertation (Maeir 1997a), written under the supervision of Profs. A. Mazar and J. Yellin. I would like to thank them both for their assistance. I would also like to thank Drs. A. Faust and E. Marcus for reading and commenting on an earlier version of this paper. Needless to say, as always, the reponsibility for any mistakes lies solely with the author. This research was assisted by grants from the Krautheimer Chair in Biblical Archaeology, Bar Ilan University, the Berman Center for Biblical Archaeology, Hebrew University, and the Rachel Yanait Ben-Zvi Prize.

2 It should be noted that other studies, while focusing on the rural components of MB II society, have stressed their relationship with the urban settlements of the period, e.g. Gophna and Beck 1981.

3 Thompson (1979; 1999:125-126) has voiced a similar opinion, claiming that there were no cities in the Bronze and Iron Age Levant. The following rebuttal applies to this opinion as well.

4 Note that in more recent studies, Falconer seems to have moderated his views and now recognizes the existence of an urban framework in the MB II southern Levant (e.g., Falconer 1995; Falconer and Savage 1995).

5 Dessel (1999), for example, who stresses the rural component during the Late Bronze Age and Iron I, does not negate the existence of urban sites during the MB II and Late Bronze Age. Note, however, that the urban character of many of the Iron age II sites has also been questioned (e.g., Ahlström 1982; London 1992).

6 The distinctions between town and village are not always self-evident, as can be seen also in the Medieval and early Modern periods in the Middle East (e.g., Lapidus 1969; Baer 1982).

7 These include: A small temple at Givat Sharett, near Beth Shemesh (Bahat 1993); A temple and at least three structures in which cult-related paraphernalia were found at the Rephaim Valley sites (Nahal Rephaim and Manahat), near Jerusalem (Edelstein, Milevski and Aurant 1998:15-16; Eisenberg 1986:56; 1991:150-151; 1993b:91-92); and an apparent MB IIA temple from Kefar Shemaryahu (Kaplan 1971:305, Fig. 11; Van den Brink 2000:46-47). To these examples can be added the MB II temple at Nahariya (Ben-Dor 1950; Dothan 1956), which seems to be in a class of its own. Although not built in an urban setting, it also was not part of a rural site, and seems to have no direct connection to any of the MB II settlements in the vicinity. It can be suggested that it in fact belongs to a third class of ritual site, that of the rural temple, which was not located within a settlement site. A similar class of temples found in the ancient Cypriote landscape has been defined and dicussed and it has been suggested that these may be an analogy of the biblical *bamah* (Wright 1992; see also Knapp 1993:58-59). Knapp (1999) has recently stressed the importance of the Cypriote rural temples within the general cultural, social and economic framework (for a critical discussion of these and other MB II temples, including questioning the cultic identification of some of them, see Katz 2000).

8 It should be stressed that the appearance of similar elements in the cultic (or, for that matter, any cultural) matrices of the urban and rural spheres (which in most cases can be classified as representing, respectively, the elite and subordinate segments of society) should not necessarily be seen as evidence of direct political and cultural domination. In fact, various similarities can be the result of attempts by the "popular" to emulate the "elite" culture. To a large extent, however, emulation is no more than an indirect facet of subordination (see, e.g., Baines and Yoffee 2000).

9 The limited accessibility of weapons during the MB II has rightfully been contrasted with the wide availability of metal weapons during the preceding MB I period. The more egalitarian social structure of the latter period may have enabled a less-restricted access to these pretige items (Ilan 1995b).

10 The existence of such ideological frameworks in the MB II has been suggested in the past (e.g., Ilan 1992:263); this suggestion is further strengthened by the many parallels from numerous ancient and modern societies (e.g., Knapp 1988; Dillehay 1990; Trigger 1990; Kolb 1994). Nakhai (2001:112-115) has noted the clear differentiation between the rural and urban cult during the

Middle Bronze II. Although, as opposed to this study, she in fact stresses the differences between the ideological realms of these rural and urban religious frameworks, it nevertheless does support our contention that the various components of the "sacred landscape" of this period mirror the urban/rural functional division of this time.

11 Nevertheless, one must take into account the possibility that such sites may have served as foci of "popular cults," which, to a certain extent, might have deviated from the norms of the "official" cults practiced in the larger urban sites. Examples of popular "dissent" over dominant ideologies, often manifested in quite subtle means, are well known from various cultures (e.g. Scott 1990). The definitions and differences of possibly similar "popular" and "official" cults in the Iron Age II have been the focus of numerous studies (e.g., Dever 1995:40-48; McNutt 1999:176-177; Miller 2000:62-105; Daviau 2001). It can be assumed that much of these discussions are relevant to the Middle and Late Bronze Ages as well, particularly in light of the evidence for continuity in the "popular" religious traditions between the Bronze and Iron Ages (see, e.g., van der Toorn 1996).

12 A possible analogy can be drawn from Ingold (1993), who argues that the "temporality" of the landscape and its features should be emphasized, and that the specific elements of the human landscape cannot be disengaged from their connection to the overall components of such a landscape. Thus, in the case under discussion, the rural/urban relationship in the MB II landscape should be examined within the wider context of the southern Levant in general (and not limited to comparing site-sizes in two disparate cultures).

REFERENCES

Ahlström, G.W.
1982 Where Did the Israelites Live? *Journal of Near Eastern Studies* 42:133-138.

Alcock, S.E.
1993 *Graecia Capta: The Landscapes of Roman Greece*. Cambridge: Cambridge University.

Baer, G.
1982 *Fellah and Townsman in the Middle East: Studies in Social History*. London: Frank Cass.

Bahat, D.
1993 Givat Sharett. Pp. 253-251 in *The New Encyclopedia of Archaeological Excavations in the Holy Land*, Vol. 1, eds. E. Stern and A. Lewinson-Gilboa. Jerusalem: Israel Exploration Society.

Baines, J. and N. Yoffee
2000 Order, Legitimacy, and Wealth: Setting the Terms. Pp. 13-17 in *Order, Legitimacy, and Wealth in Ancient States*, eds. J. Richards and M. Van Buren. New Directions in Archaeology. Cambridge: Cambridge University Press.

Ben-Dor, I.
1950 A Middle Bronze Age Temple at Nahariya. *Quarterly of the Department of Antiquities of Palestine* 14:1-41.

Blanton, R.E.
1976 Anthropological Studies of Cities. *Annual Review of Anthropology* 5:249-264.

Bunimovitz, S.
1989 *The Land of Israel in the Late Bronze Age: A Case Study of Socio-Cultural Change in a Complex Society*. Unpublished Ph.D. Dissertation, Tel Aviv University (Hebrew with English summary).

1992a The Beginning of the Late Bronze Age in Palestine: Problems in Demarcation and Terminology. *Eretz Israel* 23:21-25 (Hebrew with English summary).

1992b The Middle Bronze Age Fortifications in Palestine as a Social Phenomenon. *Tel Aviv* 19:221-234.

Christaller, W.
1966 *Central Places in Southern Germany*. Trans. C.W. Baskin. Englewood Cliffs, NJ: Prentice-Hall.

Daviau, P.M.M.
2001 Family Religion: Evidence for the Paraphernalia of the Domestic Cult. Pp. 199-229 in *The World of the Aramaeans II*, eds. P.M.M. Daviau, J.W. Wevers, and M. Weigl. Journal for the Study of the Old Testament, Supplement Series 325. Sheffield: Sheffield Academic Press.

Dessel, J.P.
1999 Tell `Ein Zippori and the Lower Galilee in the Late Bronze and Iron Ages: A Village Perspective. Pp. 1-32 in *Galilee Through the Centuries: Confluence of Cultures*, ed. E.M. Meyers. Eisenbrauns: Winona Lake, IN.

Dever, W.G.
1987 The Middle Bronze Age: The Zenith of the Urban Canaanite Era. *Biblical Archaeologist* 50:148-177.

1993 The Rise of Complexity in the Land of Israel in the Early Second Millennium B.C.E.. Pp. 98-109 in *Biblical Archaeology Today, 1990: Proceedings of the Second International Congress on Biblical Archaeology: Supplement. Pre-Congress Symposium: Population, Production and Power (Jerusalem, June 1990)*,

eds. A. Biran and J. Aviram. Jerusalem: Israel Exploration Society.
1995 "Will the Real Israel Please Stand Up?" Part II: Archaeology and the Religions of Ancient Israel. *Bulletin of the American Schools of Oriental Research* 298:37-58.

Dillehay, T.D.
1990 Mapuche Ceremonial Landscapes, Social Recruitment and Resource Rights. *World Archaeology* 22:223-241.

Dossin, G.
1970 La route de l'étain en Mésopotamie au temps de Zimri-Lin. *Revue d'Assyriologie et d'Archéologie Orientale* 64:97-106.

Dothan, M.
1956 The Excavation at Nahariyah: Preliminary Report (Seasons 1954/55). *Israel Exploration Journal* 6:14-25.

Edelstein, G., I. Milevski and S. Aurant
1998 *The Rephaim Valley Project. Villages, Terraces and Stone Mounds: Excavations at Manahat, Jerusalem, 1987-1989.* IAA Reports 3. Jerusalem: Israel Antiquities Authority.

Eisenberg, E.
1986 Emeq Refa'im- 1985. *Excavations and Surveys in Israel* 4:54-56.
1991 Emeq Refa'im- 1989/1990. *Excavations and Surveys in Israel* 9:150-56.
1993a Kitan, Tel. Pp. 878-881 in *The New Encyclopedia of Archaeological Excavations in the Holy Land 3*, eds. E. Stern and A. Lewinson-Gilboa. Jerusalem: Israel Exploration Society.
1993b Nahal Rephaim - A Bronze Age Village in Southwestern Jerusalem. *Qadmoniot* 103-104:82-95 (Hebrew).

Falconer, S.E.
1987a *Heartland of Villages: Reconsidering Early Urbanism in the Southern Levant.* Unpublished Ph.D. Dissertation, University of Arizona, Tucson.
1987b Village Pottery Production and Exchange: A Jordan Valley Perspective. Pp. 251-259 in *Studies in the History and Archaeology of Jordan 3*, ed. A. Hadidi. London: Routledge and Kegan Paul.
1995 Rural Responses to Early Urbanism: Bronze Age Household and Village Economy at Tell el-Hayyat, Jordan. *Journal of Field Archaeology* 22:349-419.

Falconer, S.E., and S.H. Savage
1995 Heartlands and Hinterlands: Alternative Trajectories of Early Urbanization in Mesopotamia and the Southern Levant. *American Antiquity* 60:37-58.

Finkelstein, I.
1992 Middle Bronze Age 'Fortifications': A Reflection of Social Organization and Political Formations. *Tel Aviv* 19:201-220.

Fuller, C.
1988 The Hindu Temple and Indian Society. Pp. 49-66 in *Temple in Society*, ed. M. V. Fox. Winona Lake: Eisenbrauns.

Gophna, R.
1979 A Middle Bronze Age II Village on the Jordan Valley. *Tel Aviv* 6:28-33.

Gophna, R. and P. Beck
1981 The Rural Aspect of the Settlement Pattern of the Coastal Plain in the Middle Bronze Age II. *Tel Aviv* 8:45-80.

Graffam, G.
1992 Beyond State Collapse: Rural History, Raised Fields, and Pastoralism in the South Andes. *American Anthropologist* 94:882-904.

Grossman, D.
1994 *Expansion and Desertion: The Arab Village and Its Offshoots in Ottoman Palestine.* Jerusalem: Yad Izhaq Ben-Zvi (Hebrew).

Haran, M.
1978 *Temples and Temple-Service in Ancient Israel: An Inquiry into Biblical Cult Phenomena and the Historical Setting of the Priestly School.* Oxford: Clarendon Press.

Hodder, I. and C. Orton
1976 *Spatial Analysis in Archaeology. New Studies in Archaeology.* Cambridge: Cambridge University.

Holladay, J.S., Jr.
2001 Toward a Paradigmatic Understanding of Long-Distance Trade in the Ancient Near East: From the Middle Bronze II to Early Iron II - a Sketch. Pp. 136-198 in *The World of the Aramaeans: Biblical Studies in Honour of Paul-Eugène Dion II*, eds. P.M.M. Daviau, J.W. Wevers, and M. Weigl. Journal for the Study of the Old Testament, Supplement Series 325. Sheffield: Sheffield Academic Press.

Ilan, D.
1992 A Middle Bronze Age Offering Deposit from Tel Dan. *Tel Aviv* 19:247-266.
1995a The Dawn of Internationalism: The Middle Bronze Age. Pp. 297-319 in *The Archaeology of Society in the Holy Land*, ed. T.E. Levy. London: Leicester University.
1995b Mortuary Practices at Tel Dan in the Middle Bronze Age: A Reflection of Canaanite Society and Ideology. In *The Archaeology of Death in the Ancient Near East*, eds. S. Campbell and A. Green. Oxbow Monograph 51. Oxford: Oxbow.

Ingold, T.
1993 The Temporality of Landscape. *World Archaeology* 25:152-174.

Joffe, A.H.
1991 Early Bronze I and the Evolution of Social Complexity in the Southern Levant. *Journal of Mediterranean Archaeology* 4:3-58.
1993 *Settlement and Society in the Early Bronze I and II Southern Levant: Complementarity and Contradiction in a Small-Scale Complex Society.* Monographs in Mediterranean Archaeology 4. Sheffield: Sheffield Academic Press.

Kaplan, J.
1971 Mesopotamian Elements in the Middle Bronze II Culture of Palestine. *Journal of Near Eastern Studies* 30:293-307.

Katz, J.C.
2000 *The Archaeology of Cult in Middle Bronze Age Canaan: The Sacred Area of Tel Haror, Israel.* Unpublished Ph.D. Dissertation, University of Pennsylvania, Philadelphia.

Knapp, A.B.
1988 Ideology, Archaeology and Polity. *Man (New Series)* 23:133-163.
1993 *Society and Polity at Bronze Age Pella: An Annales Perspective.* JSOT/ASOR Monograph Series 6. Sheffield: Sheffield Academic Press.
1997 *The Archaeology of Late Bronze Age Cypriot Society: The Study of Settlement, Survey, and Landscape.* Department of Archaeology Occasional Paper 4. Glasgow: University of Glasgow.
1999 Ideational and Industrial Landscape on Prehistoric Cyprus. Pp. 229-251 in *Archaeologies of Landscape: Contemporary Perspectives*, eds. W. Ashmore and A.B. Knapp. Social Archaeology. London: Blackwell.

Kolb, M.J.
1994 Monumentality and Rise of Religious Authority in Precontact Hawai'i. *Current Anthropology* 34:521-547.

Lapidus, I.M.
1969 Muslim Cities and Islamic Societies. Pp. 47-74 in *Middle Eastern Cities*, ed. I.M. Lapidus. Berkeley: University of California.

London, G.A.
1992 Tells: City Center or Home? *Eretz Israel* 23:71*-79*.

Mabry, J.B.
1985 *The Socioeconomic Geography of Agrarian City-States in the Middle Bronze Age II of Palestine: The Urban-Rural Hierarchy.* Unpublished paper on file.

Maeir, A.M.
1997a *The Material Culture of the Central and Northern Jordan Valley in the Middle Bronze Age II: Pottery and Settlement Pattern.* Diss. Jerusalem: Hebrew University.
1997b T.1181, Area L, Hazor: A Multiple Interment Burial Cave of the MB IIa/b Period. Pp. 295-340 in *Hazor V*, ed. by A. Ben-Tor. Jerusalem.
1998 A Computer Simulation of Middle Bronze II Site Territories: The Jordan Valley and Beyond. Conference presentation. In *The First International Conference of Ancient Near Eastern Archaeology*, Rome, May, 1998.
2000a The Political and Economic Status of MB II Hazor and MB II Trade: An Inter- and Intra-Regional View. *Palestine Exploration Quarterly* 132:1-22.
2000b Hamadiyah-North: A Rural MB II Settlement in the Beth Shean Valley. *'Atiqot* 39:31-42.

Magness-Gardiner, B.
1994 Urban-Rural Relations in Bronze Age Syria: Evidence from Alalah Level VII Palace Archives. Pp. 37-47 in *Archaeological Views from the Countryside: Village Communities in Early Complex Societies*, eds. G.M. Schwartz and S.E. Falconer. Washington, D.C.: Smithsonian Institution.

Magness-Gardiner, B., and S.E. Falconer
1994 Community, Polity and Temple in a Middle Bronze Age Levantine Village. *Journal of Mediterranean Archaeology* 7:127-164.

Mazar, A.
1990 *Archaeology of the Land of the Bible: 10,000-586 B.C.E. The Anchor Bible Reference Library.* New York: Doubleday.
1991 Sanctuaires et Temples en Canaan. Pp. 1258-1286 in *Supplement au Dictionnaire de la Bible 11*. Paris: Letouzey & Ane.
1999 The 1997-1998 Excavations at Tel Rehov: Preliminary Report. *Israel Exploration Journal* 39:1-42.

McIntosh, S.K. and R.J. McIntosh
1993 Cities with Citadels: Understanding Urban Origins along the Middle Niger. Pp. 622-641 in *The Archaeology of Africa: Food, Metals and Towns*, eds. T. Shaw et al. One World Archaeology 20. London: Routledge.

McNutt, P.
1999 *Reconstructing the Society of Ancient Israel. Library of Ancient Israel.* Louisville: Westminster John Knox.

Miller, P.D.
2000 *The Religion of Ancient Israel. The Library of Ancient Israel.* Knoxville: Westminster John Knox.

Moser, C., and W. Scott
1961 *British Towns: A Statistical Study of of Their Social and Economic Differences.* Edinburgh: Oliver and Boyd.

Nakhai, B.A.
2001 *Archaeology and the Religions of Canaan and Israel.* ASOR Books 7. Boston: American Schools of Oriental Research.

O'Connor, D.
1991 Early States Along the Nubian Nile. Pp. 145-165 in *Egypt and Africa: Nubia from Prehistory to Islam*, ed. W.V. Davies. London: British Museum.

Palumbo, G.
1990 *The Early Bronze Age IV in the Southern Levant: Settlement Patterns, Economy, and Material Culture of a 'Dark Age'.* Contributi e Materiali di Archaeologia Orientale III. Rome: Universitá Degli Studi di Roma La Sapienza.

Philip, G.
1991 Cypriot Bronzework in the Levantine World: Conservatism, Innovation and Social Change. *Journal of Mediterranean Archaeology* 4:59-107.

Pollock, S.
1999 *Ancient Mesopotamia: The Eden That Never Was.* Cambridge: Cambridge University Press.

Rappaport, R.A.
1999 *Ritual and Religion in the Making of Humanity.* Cambridge: Cambridge University Press.

Renfrew, A.C.
1994 The Archaeology of Religion. Pp. 47-54 in *The Ancient Mind: Elements of Cognitive Archaeology*, eds. A.C. Renfrew and E.B.W. Zubrow. New Directions in Archaeology. Cambridge: Cambridge University Press.

Renfrew, A.C. and E.V. Level
1979 Exploring Dominance: Predicting Polities from Centers. Pp. 145-167 in *Transformations: Mathematical Approaches to Culture*, eds. A.C. Renfrew and K.L. Cooke. New York: Academic Press.

Scott, J.
1990 *Domination and the Arts of Resistance: Hidden Transcripts.* New Haven: Yale University Press.

Smith, C.A.
1976 Regional Economic Systems: Lining Geographic Models and Socioeconomic Problems. Pp. 3-63 in *Regional Analysis I: Economic Systems*, ed. C.A. Smith. Studies in Anthropology. New York: Academic.

Smith, M.S.
1989 Cities, Towns, and Urbanism: Comment on Sanders and Webster. *American Anthropologist* 91:454-460.

Thompson, T.L.
1979 *The Settlement of Palestine in the Bronze Age.* Beihefte Zum Tubinger Atlas Des Vorderen Orients, Reihe B Nr. 34. Wiesbaden: L. Reichert.
1999 *The Mythic Past: Biblical Archaeology and the Myth of Israel.* London: Basic Books.

Trigger, B.G.
1990 Monumental Architecture: A Thermodynamic Explanation of Symbolic Behaviour. *World Archaeology* 22:119-132.

Van den Brink, E.C.M.
2000 A Middle Bronze Age IIA Burial Pit at Kefar Shemaryahu. *'Atiqot* 39: 43-47.

van der Toorn, K.
1996 *Family Religion in Babylonia, Syria and Israel: Continuity and Change in the Forms of Religious Life.* Leiden: Brill

Wells, P.S.
1984 *Farms, Villages and Cities: Commerce and Urban Origins in Late Prehistoric Europe.* Ithaca: Cornell University.
1994 Changing Models of Settlement, Economy, and Ritual Activity: Recent Research in Late Prehistoric Central Europe. *Journal of Archaeological Research* 2:135-163.

Wright, G.R.H.
1992 The Cypriot Rural Sanctuary. An Illuminating Document in Comparative Religion. Pp. 269-283 in *Studies in Honour of Vassos Karageorghis*, ed. G.C. Ioannides. Nicosia: Etaria Kypriakon Spoudon.

Zaccagnini, C.
1989 Asiatic Mode of Production and Ancient Near East: Notes Towards a Discussion. Pp. 1-126 in *Production and Consumption in the Ancient Near East*, ed. C. Zaccagnini. Budapest: University of Budapest.

The 10th century B.C. Settlement of the Negev Highlands and Iron Age Rural Palestine[1]

MORDECHAI HAIMAN

Israel Antiquities Authority
P.O.B. 586
Jerusalem, 91905
Israel
moti@israntique.org.il

Approximately 350 sites dating to the Iron Age are known from the Negev Highlands, most of which can be dated to a rather brief time, during the tenth century BCE. Following a review of the relevant data, these settlements are compared to the rural sites in the more northern regions of Palestine, sites that began to appear in the late eleventh century BCE. Based on the similarities between the sites in these two regions, it is suggested that the impetus for the founding of the settlements of the Negev Highlands should be understood within the context of the social and geo-political situation during the time of the Israelite United Monarchy. Once the United Monarchy was divided and the geo-political conditions in the region shifted, the settlements in the Negev Highlands lost their justification and were abandoned.

Approximately 350 Iron Age II sites are known in the Negev Highlands, (Woolley and Lawrence 1915:61-71; Anati 1955; Glueck 1959; Evenari et al. 1958; Aharoni et. al. 1960; Aharoni 1967; 1976; Rothenberg 1967:71-9; Meshel 1974; 1994; Cohen 1981b; Haiman 1986; 1991; 1993; 1998; in press; Lender 1990). The sites are comprised of fortresses, dwelling structures, and installations, such as water cisterns, threshing floors and silos. Due to the desert conditions, the sites were found in an excellent state of preservation, with most of the elements intact.

Examination of old and recent publications of surveys and studies dealing with Iron Age rural Palestine (Kochavi 1972; Eitam 1980; Dar 1986; Finkelstein 1988; Zertal 1988; Maitlis 1989; Faust 2000) shows that a similar type of rural settlement was distributed throughout the entire sedentary lands north of the Negev, constituting the main component of settlement in inter-city and village areas, especially in peripheral regions, such as the Samarian and Judaean deserts.

The Architecture and Structure of the Settlement in the Negev

The 350 Iron Age sites presently known in the Negev Highlands include 62 fortresses, 1,200 dwelling structures, and 360 animal pens.

Fortresses

Approximately 60 casemate fortresses and square towers have been found to date in the the Negev.

Casemate Fortresses: Oval or round in shape, the diameters of these fortresses range from 15-50 m. (see figures 5-8-10). A few fortresses are square or rectangular, ranging in size between 18 X 20 and 20 X 22 m. (see figures 5-8-10). This type of fortress has been discussed extensively in the literature (Aharoni 1957; Cohen 1980; Herzog 1983; Finkelstein 1984; Eitam 1988; Meshel 1994).

Square Towers: These towers range from 5 x 7 m. – 10 x 12 m., and are built of large stones up to 1.5 m. long. On the inside, they usually contain between one and four casemate rooms surrounding a central courtyard, but a few do not have casemate rooms at all (See figure 5-2 and 5-11). Most of the square towers are single structures, while the remainder are installed in the walls of casemate fortresses or attached by round animal pens. In some cases, square towers were even found in the courtyards of casemate fortresses. The dimensions and features of the towers are similar to those of the four-room house, the main difference being that the rooms of the towers are divided by walls, while those of the four-room house are divided by rows of columns, sometimes installed in the walls (Haiman 1986:17).

Structures

Hundreds of dwelling structures were surveyed in the vicinity of the fortresses. These structures can be divided into three groups:

Group I: Solid structures designed like those in the northern sedentary lands, consisting of 1-4 elongated rectangular rooms and columns. The structures are built of large stones, 0.60 - 1 m. long, and some even longer (see figure 5-3:1-4 and figure 5-12). Most of the structures in this group were located near water cisterns, and it can be assumed that they were permanent. Of the total number of 1195 dwelling structures, 140 (11%) belong to this group, and 30 of these are of the four-room house type (Cohen 1986: Sites 48, 67, 88, 120; Haiman 1986: Sites 50, 103, 133, 172; Lender 1990: Site 243).

G r o u p II: Structures consisting of one or more rooms, the architectural style of which is similar to Group I, with additional round rooms and courtyards or animal pens (Aharoni et al. 1960: Figs 8-10; Lender 1990: Site 55; see here fig 5-3:5-8). The group also includes a type of enclosed structure containing dwelling rooms, and even entire structures, built around a central round courtyard (Aharoni et al. 1960: Figure 5; Haiman 1986: Site 271; 1991: Site 1; Cohen 1981b: Site 17; see here fig 5-3:10). These enclosed structures reflect an architectural tradition characteristic of the desert region, and not the architecture of the fortresses, as has been suggested (Herzog 1984:41-9; Finkelstein 1988: 238-59). Group II consists of 177 structures, which can also be considered permanent structures.

G r o u p III: One-room structures of various shapes -- rectangular, round, oval and horseshoe -- measuring 2-3 x 4-5 m. The walls are built of small stones, 0.20 - 0.40 m. long, with some larger, mainly in the doorways (see figure 5-3:9). The shallow debris suggests that most of the structures were huts. Since the majority of the settlements with structures of this type did not correspond with a water source, they were probably used for seasonal habitation (Haiman 1986: Sites 236, 237). A total of 878 structures belong to this group.

Settlements

Only a few of the sites in the Negev Highlands can be defined as villages (Cohen 1981b: Sites 64, 98), resembling those in the north (Faust 2000). In most cases, the settlements are characterized by structures scattered over a wide area, which raises the problem of how to define an actual settlement. It seems that the criteria should be the existence of agricultural installations and water cisterns in the vicinity of the structures (see figure 5-4). Accordingly, 220 out of 350 sites from the Iron Age can be defined as settlements. Of these 220 sites, 13 contained one structure each, 133 contained 2-6 structures, 59 contained 7-11 structures, and 15 contained 12-40 structures. Those with 2-6 structures apparently constitute a functional settlement characteristic of the Negev Highlands. The larger groups probably combined a number of settlements in the vicinity.

Another question concerning the issue of defining a settlement is the correspondence between sites and fortresses. In many cases, neighboring fortresses and settlements constitute a settlement unit, for example Atar ha-Roah and Horvat Ramat Boqer. (Cohen 1981b: site 64, Cohen 1985: Site 48). However, such a correspondence (see figs. 5-5 and 5-6) is not consistent. Numerous settlements are 3-4 km. away from a fortress and the relationship between them is not readily apparent. Other fortresses are not located near settlements. In any case, almost all of the sites are scattered within the general distribution areas of the fortresses and not outside of these areas.

The fact that the sites are well preserved enables us to estimate the size of the population on the basis of the number of structures and the number and area of dwelling rooms. It can be assumed that the structures defined as permanent accommodated families with an average of 4-6 members. The total population living in 317 permanent structures can be roughly estimated at 1500 individuals. Based on the living space in these structures (790 dwelling rooms totaling 12,500 sq. m.), the population can be estimated at 1,250 (Naroll 1962; Shiloh 1980; Broshi and Gophna 1984; Cohen 1986:362; Finkelstein 1988:330-35).

The 878 structures of Group III are small and poorly built, and could not contain an entire family, but potentially could have accommodated one or two people and may have been used for seasonal agriculture or animal husbandry activities conducted at some distance from a permanent settlement. The number of people accommodated in these structures can be estimated at 850, based on the calculation of 878 dwelling rooms with a total area of 8,500 sq. m.

In 58 fortresses and square towers, there were 450 rooms with a total area of 11,250 sq. m., which could accommodate around 1,100 people.

Altogether, the dwelling rooms at the sites could accommodate as many as 3,500 people, but the actual population was probably smaller. It can be assumed that only the permanent sites reflect the size of the population, since the temporary sites were used seasonally by the population of the permanent settlements. Similarly, it can be assumed that the fortresses did not serve as regular residential settlements. Analysis of the data published from excavations in the fortresses (Cohen 1986) attests that only 60 percent of the rooms had any evidence of occupation and the population capacity in most of the fortresses is low (Haiman 1988: 98-101). It is assumed that the fortresses served mainly as observation points for small groups of soldiers patrolling the area. It is noteworthy that the dwelling area in the permanent settlements was similar to that of the fortresses; thus, in times of danger, people from the settlements could find refuge in the fortresses. In general, the population of the settlements was about 1500, with an unspecified number of people living in the fortresses.

Settlement Distribution, Spatial Relationships and Environment

Settlement in the Negev Highlands consisted of small sites and individual structures within an area visible from the fortresses, but at far distance from the natural water sources. From the viewpoint of desert environmental conditions, the phenomenon of a distribution of sites in the center of the Negev Highlands, at a distance of more than 10 kilometers from a water source, is unique to the Iron Age. In contrast, in the Early Bronze Age II and Middle Bronze Age I, large permanent sites were built only in proximity to water sources. (Haiman 1989:179-183). Although cisterns were dug, they could not sustain a permanent population for any length of time. The area also has low potential for a livelihood based on agriculture and animal husbandry. The average annual rainfall ranges from 80-100 mm.; and considering the numerous draught years, the potential for cultivating the land is minimal.

Around 80% of the dwelling sites, as well as all the fortresses and square towers (with the exception of Tell el-Qudeirat), were built on mountain ridges and slopes, at points providing a wide panoramic view. Most of the fortresses were built on the tops of steep slopes and are difficult to reach. Even those built at relatively low points, such as Atar ha-Roah and Horvat Haluqim, provide a broad view of the surroundings, to a distance of several kilometers. Many of the fortresses overlook one another and almost all have a view of the surrounding settlements. Almost all the fortresses are located slightly below the ridge line of the slopes, concealing them from view from below.

Examination of the distribution of sites shows that there was a main distribution line and two secondary lines. The main line runs along the Dimona-Mishor Yeroham - Sede Boqer - Ramat Matred - Ramat Barnea route, cutting through the heart of the Negev Highlands on a flat surface axis, 100 km. long and 2-8 km. wide (see fig 5-1). The line is easily accessible, with fortresses overlooking the entire area, and "ends" near Quseima, at the westernmost known fortress (Meshel 1994). It has been identified as the biblical *Derekh ha-'Atarim* (Num. 21:1) (Aharoni et al. 1960:24).

One of the two secondary distribution lines diverges from the main line in the area of Har Nafha, and continues southward along the edge of Makhtesh Ramon, westward to the fortress of Nahal Lotz, and then north along Nahal `Aqrab (see fig 5-1). The other secondary line leads north from Sede Boqer, along Nahal ha-Besor, towards Be'erotayim and Quseima.

There is evidence of fluctuation in the number and size of structures and settlements in the various areas, as a result of environmental conditions, with fewer permanent and more temporary structures as one progresses southward. Between Dimona and `Avdat, most structures are permanent, with an average of six structures at each site; 25 of the 30 four-room houses known in the Negev Highlands were found in this area. At Ramat Matred and Ramat Barnea, 246 sites were surveyed, of which 45 can be defined as permanent. The sites contained a total of 109 permanent structures with an average of two to three per settlement. These permanent structures are smaller and contain fewer rooms than those in the Dimona-`Avdat area. In the Mishor ha-Ruhot-Lotz area a larger number of permanent sites were found, but these contain one or two-rooms structures with the exception of one four-room house found at Ma'agorot Lotz.

In the vicinity of 'Ein el-Qudeirat, the majority of settlements can also be defined as permanent, although they contain small structures, mainly with one or two rooms. No such significant differences were found between the fortresses of the various regions, and no difference was evident in the solidity of their construction from area to area.

Approximately 80% of the sites at Ramat Matred and Ramat Barnea are temporary. They are scattered over a large area, and were used for agricultural activity or sheep and goat husbandry. Most of the remaining sites, which can be considered permanent, include a variety of structures, while the permanent sites in the Dimona-`Avdat area are more homogeneous. Concentrations of sites and structures, for example, in Nahal Mitnan and Nahal Horesha, demonstrate the characteristics of settlement in the southern distribution areas. One characteristic is the lack of uniformity among the buildings in numerous settlements, which include a combination of permanent structures, enclosed structures and temporary structures. It can be assumed that the sites were established over several years, as groups of structures were added from time to time, reflecting the semi-nomadic lifestyle of most of the population.

Desert Livelihood in Practice

Water supply: The most important factor for survival in the desert is water. As shown above, however, the distribution of settlements was not affected by the locations of natural water sources and the population was almost entirely dependent on water cisterns. Two types of water cisterns are dated to the period. One is a rock-cut cistern of which dozens of them were found in proximity to sites and fortresses. The other type is an open reservoir, 3-20 m. in diameter and 2-5 m. deep, dug into sealed earth (see figure 5-13). The walls were paved with courses of stones and no traces of plaster were found (Evenari et al. 1958:241- 43; Moran and Palmach 1985; Cohen 1986:348-53).

Over 100 reservoirs were found in association with the distribution area of the Iron Age sites, most of them along the Yeroham - Sede Boqer - Mishor ha-Ruhot - Ma'agorot Lotz line. It was found that the greater the distance of settlements from natural water sources, the larger the ratio of reservoirs to permanent structures. In the Sede Boqer area, located 7 km. from a natural water source, there were 7 permanent structures per reservoir, and the average amount of water available for each structure dweller (=family) was approximately 75 cubic m. In the Ramat Matred and Ramat Barnea areas, located 10 km away from a natural water source, there were around 4 small permanent structures per reservoir (100 cu. m. per structure). In the Mishor ha-Ruhot-Ma'agorot Lotz area, which is farthest away from a natural water source, three reservoirs were found per permanent structure. In the `Ein el-Qudeirat area, there was only one reservoir for 38 permanent structures.

Theoretically, the 100 reservoirs alone, regardless of the rock cut cisterns, could provide considerably more than the annual water consumption required for the population reflected in the sites- about 18co/m for a family and its animals (Evenari, Shanan and Tadmor 1971:148-171). However, the evaporation rate in the desert, which is around 2.5 m. per year, raises questions about the capability of the reservoirs to hold water beyond the beginning of the summer. Taking into consideration the frequent droughts (every 2-3 years), the desert phenomenon of localized rainfall and considerable differences in the amount of annual rainfall from one year to the next, the cisterns of both types were not filled, or not filled to full capacity every year. It is clear that the reservoirs could not have solved the problem of sufficient water supply for permanent settlement, but may reflect a tradition brought in from the outside. The dependence on cistern water in the conditions of the desert may reflect, as a whole, an unstable population (Negev 1979).

Pastoralism: Findings show that the inhabitants were engaged in sheep and goat husbandry. A total of 360 animal pens were found covering an area of 45,000 sq. m., indicating the central role of sheep and goat husbandry in the economy. In order to calculate the size of the herds, it is assumed that only the pens located in the permanent settlements reflect this number (as in estimating the size of the population). A total of 100 animal pens covering an area of 13,000 sq. m. were found in the permanent settlements. Assuming that 2 sq. m. are required per head (Zuk 1991), there would have been a total of 6,500 animals in the permanent settlements. On average, the area of the animal pens was 40 sq. m. per dwelling structure, indicating a herd of 20 animals per family, which is close to the size of the average herd of present-day Bedouin families in the southern Negev and the Sinai Peninsula (Evenari, Shanan and Tadmor 1971:148-50; Baily 1977:242-43).

Agriculture: Considerable evidence of agricultural activity was also found: an abundance of sickle blades, silos (similar to those found in the north [Finkelstein 1988:264-69]) and threshing floors. Around 80 silos were found in the area surveyed by the author, which can be classified into two types: a) roofed, bell-shaped silos that widen towards the bottom, with the bases ranging between 1-2 m. in diameter. The upper part is around 1 m. in diameter, and the depth is approximately 2 m. (see figure 5-14); and b) open, round or square silos, 1-2 x 1-4 m. in area and 1-2 m. deep (see figure 5-15). Both types are dug into the ground and lined by small stones.

Thirty threshing floors attributed to Iron Age sites in the area surveyed by the author are comprised of a circle of stones, 8-15 m. in diameter, surrounding a leveled comprise-dirt floor. Ten of the threshing floors are abutted by silos (see photograph 9), indicating that the grain was stored immediately after threshing.

A related question is the availability of agricultural land, since the only arable land in the area is the wadi bed. Although terraced wadis have been dated to this period, surveys show that many Iron Age sites at which the above-mentioned evidence of agricultural activity is found, were not located near terraced wadis. Furthermore, in many areas, such as Shivta, where most of the wadis were terraced, Iron Age settlement is sparse. The terraced wadis in the Negev encompass ca. 300,000 dunams, while less than 4,000 dunams would have been required to grow enough grain to sustain the Iron Age population (Bruins 1985:86-95; Rosen 1986: 171-74). Moreover, there is no correspondence between the sophisticated agriculture reflected in the terraced wadi system and the low level of the material culture represented at the Iron Age sites. The excavations of terraced wadis conducted at Wadi el Qudeirat, Nahal Mitnan and Horvat Haluqim showed that the wadis were terraced only after the Iron Age (Goldberg 1984; Bruins 1985). Taking into account changes in the amount of natural soil in the wadi beds, it appears that, at least until the Iron Age, there was enough natural soil for agricultural cultivation and large-scale terracing of the wadis was not necessary. It seems that the problem did not lie in the amount of land available for cultivation, but rather in growing grain in the desert environment, with its average of 80 mm. annual rain fall.

Taking into consideration the severe limitations in water supply and the frequency of draught, it is clear that the inhabitants of the Negev could not support themselves. The conclusion, therefore, is that other considerations, not connected with the exploitation of environmental conditions for subsistence, caused this population of farmers and shepherds to migrate into the desert.

Rural Palestine and the Origin of the Settlement in the Negev

At the end of the eleventh century B.C.E., a new type of rural settlement emerged, comprised of a small group of buildings linked to a fortress with relatively thin walls and installations such as water cisterns and silos. This type of settlement continued until the end of the Iron Age and is most appropriately described by the biblical term "Haserim" (Gophna and Singer-Avitz 1984). The rural settlement was distributed throughout the northern sedentary lands, in inter-city areas, and in peripheral regions, such as the Samarian and Judaean deserts (see figure 5-7). In a few cases, such settlements were built above sites from earlier stages of the Iron Age. It can be assumed that the new type of settlement reflects the social and cultural changes that took place in the process of the transition from a tribal society to a monarchy. On the basis of the similarity between the settlement in the Negev Highlands and that in the north, it is suggested that the Negev settlements are an expansion from the north. Although most of the examples of rural settlements presented below are later than the Negev settlements, they may be viewed as providing a relevant typological perspective (A. Mazar 1982:97).

The Northern Negev: Significant evidence was found at the site of Tel Masos on the southern periphery of the sedentary lands. This large Iron Age I site (Strata II and III) was destroyed at the end of the eleventh century B.C.E, and on top of the ruins, a small settlement, dated to the end of the eleventh or early tenth century B.C.E. was built, comprised of a small square fortress and a number of dwelling structures and silos (Stratum I) (Fritz and Kempinski 1983).

Small settlements associated with fortresses or cities are also known to have existed in the area from the eighth-sixth centuries B.C.E., for example at Horvat `Uzza (Beit-Arieh 1986) and `Aro`er, where approximately 25 small silos of the type characteristic of the Negev Highlands were found (Biran and Cohen 1981:265, 271).

Judaean Desert: There is a substantial basis for comparison between these and the Negev Highlands sites, despite the 200-year chronological gap. The similarity is reflected in the variety of fortresses (round and rectangular-shaped and square towers), the distribution of dwelling structures and water cisterns associated with the fortresses, the frequency of round buildings and the large number of animal pens (Evenari et al. 1958:238). Of the 50 sites from the 8-6th century B.C.E. surveyed by Bar-Adon (Bar-Adon 1972: Sites 4, 21, 49, 82-83, 91, 93, 95, 186, 201-202), 13 could be classified as small settlements resembling those of the Negev. The same applies to several other Judaean Desert sites, such as Horvat Abu Tabak and Karm es-Samrah, which are comprised of fortresses with dwelling structures scattered around them, pens, water cisterns and installations (Cross and Milik 1956; Stager 1976).

Judaean Hills: Numerous fortresses and small settlements were surveyed throughout the hinterland areas between the cities (Kochavi 1972: Sites 7, 13, 28, 39, 79, 85, 166, 246, 250). Most of the sites are dated later than the Negev Highlands sites. One of the sites -- Giloh (Mazar 1994: 78-90) -- may contribute to the chronological issue: it is suggested that Tower G and a second tower in the vicinity of the site reflect the transition from a typical Iron Age I enclosed site to a site that includes a square tower, as Tel Masos.

On the western periphery of the Judaean Hills, the settlement pattern that prevailed from the eighth to the sixth centuries B.C.E. included thin-walled fortresses, 4-5 km. apart, water cisterns and small sites scattered over a large part of the immediate vicinity, resembling Negev Highlands sites, despite the chronological gap (A. Mazar 1982:97).

Samaria: The occupation of the hinterland, like that of the Judaean Hills consist of small sites including fortresses with thin walls, strategically located to overlook dwelling structures, water cisterns and agricultural installations at lower altitudes (Kallai 1972: Sites 2, 15, 31, 41, 47, 68, 69, 86, 118, 250; Gophna and Porath 1972: Sites 54, 162, 165, 174, 197, 212; Eitam 1980:64-69; Dar 1986:213-17; Finkelstein and Lederman 1997). Here too, as in the Judaean Hills settlement, there is a noticeable increase as one moves away from the center of the country towards the desert in the east or towards the western periphery (Zertal 1996:81-84; Finkelstein and Lederman 1997:131-287). The following sites are noteworthy:

Tell el-Ful: Like Tel Masos, the site reflects the appearance of the type of settlement discussed in this paper. The first fortress (Fortress No. I) was built onto a settlement at the end of the eleventh century B.C.E., with silos and agricultural installations found around the fortress (Lapp 1981; Finkelstein 1988:56-60; A. Mazar 1994:76-8).

Khirbet ed-Dawwara: A round settlement, approximately 80 m. in diameter. The site is surrounded by a solid wall with several structures attached to it, forming a kind of casemate, with several water cisterns in the courtyard (Finkelstein 1988:64). The site is roughly contemporaneous with the Negev fortresses and its shape indicates that the round contour was not an exclusive feature of the Negev.

`Izbet Sarta: During the last occupation phase of this site (Stratum I), at the end of the eleventh century B.C.E, it was reduced to a single structure with silos scattered around it. This single structure is the final use of a four-room house of the larger site from the previous period (Stratum II). In the last phase, the spaces between the columns of the four-room house were filled with stones, and an inner wall was built parallel to the northern wall (Wall J122). The structure was also expanded to include five rooms surrounding a central courtyard (Finkelstein 1988:78-80). It is suggested that the four-room house apparently was converted into a square tower similar to those found in the Negev Highlands, which lacked inner row columns.

The Galilee: In the Lower Galilee, there are traces of the transition from Iron Age I types of sites, to small rural sites in the inter-urban areas as well as a number of fortresses (Zori 1977:6, 89; Gal 1992). Horvat Talil in the Upper Galilee is a casemate fortress dating to the end of the 11th century B.C., built on top of the remains of a site dating to the Iron Age I (Aharoni 1957:18-24).

Rural Iron Age sites comprised of a fortress and nearby structures have also been surveyed in various regions in Jordan (Banning and Fawcett 1983:302; Gordon and Knauf 1987; Miller 1990; Palumbo, Marby and Kuijit 1990:99; Worschech 1990), but their inclusion in the discussion is beyond the scope of this article.

Based on the material culture and pattern of settlement, it can be assumed that most of the Negev Highlands population immigrated from the northern sedentary lands. It should be mentioned that proposals suggesting other origins disregard the material culture (Eitam 1988; Na'aman 1990; Finkelstein 1996). The differences between the Negev sites and those of the north are not substantial, except for some regional features characteristic of the desert, such as a preference for round buildings and a large number of temporary settlements. The "Negbite" pottery (40% of the pottery assemblage) and the few cairn burials may reflect the integration of the population or at least the incorporation of cultural elements of the local desert population (Meshel 1979; 1994:53-61; Cohen 1986:385-94; Haiman 1988:105-9).

In contrast, it has also been argued that the region was settled by local nomads, such as the Amalekites (Rothenberg 1967:92-101; Finkelstein 1984). This argument can be refuted on the following grounds:

1. There are no findings in the Negev indicating that local settlement developed from an early nomadic to a permanent settlement stage. Indications concerning the appearance of the settlement type discussed here were found in the north, in particular at sites like Tel Masos, 'Izbet Sarta, Tell el-Ful and Horvat Talil.

2. The findings and the architecture of the Negev Highlands sites indicate that the settlements were both established at the same time and abruptly abandoned. Had the population been local, the processes of establishment and abandonment would probably have been more gradual.

3. The prevailing view is that semi-nomads can subsist only alongside permanent settlements, as part of a wider economic network (Marx 1974; Baily 1977:242-43; Shmueli 1977:211; Khazanov 1983). One manifestation of this phenomenon is the physical proximity of semi-nomads to the peripheral areas of permanent settlements. At the end of the Late Bronze Age and beginning of the Iron Age, the southern periphery of the main permanent settlement region of Palestine was in the southern Shephelah and the southern coastal area (Na'aman 1988; Rainey 1988). During this period, the possible living area for semi-nomads was most likely in the northern Negev, where they could maintain close ties with the populations of southern Palestine (Egyptians, Philistines, Canaanites and Israelites).

4. The biblical sources use the expression "...as far as Shur, to the land of Egypt" (1Sam. 27:8) to describe the domain of the Amalekites". This applies to the northwestern Negev (Na'aman 1980) rather than to the other southern peripheries, which are characterized by expressions such as "the wilderness of Beer-sheba" (Gen.21:14) and "the wilderness of Zin" (Num. 20:1, 27:14; Deut. 32:51; Josh. 15:1).

5. As a rule, semi-nomads do not settle in permanent settlements at their own initiative. Contemporary Bedouins, for example, have settled as a result of a government initiative, or through relations with a central regime, after the vital interim stage of spontaneous settlement. The economy of nomads is minimalistic, with hardly any sources of capital and no potential for independent growth (Shmueli 1980:71-122; Khazanov 1983:198-202). Thus, nomads, living in poor regions like the Negev Highlands, could not have established a dense settlement system on their own.

6. The distribution of settlements in the Negev Highlands, as mentioned above, does not accord with the environmental conditions. The disregard for the distribution of natural water sources is particularly obvious. This indicates that there was an element interested in controlling the area, but not necessarily in living off its resources. It can be assumed that local inhabitants would settle near natural water sources, on the principle that nomads who are in a process of transition to permanent settlement would migrate to an area with better environmental conditions (Khazanov 1983:200).

Chronology and Conclusions

Excavations at dozens Negev Highlands sites of this period suggest that the settlement was of a short duration, although there are differences of opinion regarding the exact dating. The rich pottery assemblage published by Cohen (1986:363-84, 410-35) has provided a basis for dating the sites to the tenth century B.C.E. The earliest pottery is from the end of the eleventh century B.C.E., while some types do not predate the tenth century B.C.E. There is no ceramic evidence typical of the earlier part of the eleventh century B.C.E., such as Philistine pottery; rather, the coastal region is represented by later types, such as black juglets and Cypro-Phoenician pottery.

The ceramic dating corroborates references to the Negev in Shishak's list, including names like "Ezem," "Arad," and others, with the prefixes "Pa-Negeb" and "Hagar," which refer to fortresses. Shishak's list of around 80 site names refers, therefore, directly to the Negev Highlands, since it could only have been based on the short-lived settlements that existed in the region before his campaign (B. Mazar 1957; Ahituv 1979). Shishak's campaign to Palestine apparently led to the abandonment of the settlements in the Negev Highlands. Taking into account the ceramic finds and Shishak's list together with the chronological framework, the settlement in the Negev Highlands can be dated from the end of eleventh century B.C.E. to 925 B.C.E.

The origin of the type of settlement found in the Negev was in the northern sedentary lands, where it appeared at the end of the eleventh century B.C. The Negev settlement apparently represents the first use of this type of site as an element in a system of fortified settlements in a peripheral region. Subsequently, this pattern of settlement served as a model for settlement in the peripheral regions of the north.

It is suggested that the settlement in the heart of a desert, virtually lacking in sources of livelihood, was initiated and supported by the Israelite United Monarchy to protect the interests of the state in response to geopolitical conditions that threatened its southern border. The main threat was represented by Shishak's subversive activity against King Solomon, the Egyptian control of the southern coast, and that Hadad, the ruler of Edom, was Shishak's vassal (Malamat 1983:167-95; Bartlett 1980:107-13). A system of fortified frontier settlement in the Negev Highlands could have served as a territorial barrier between Egypt and Edom, in order to prevent the establishment of a joint front in the south. Control of the Negev Highlands could also have provided alternative access to the Red Sea, since the route via Edom was most probably impassable. The geopolitical conditions that could have necessitated the Israelite presence in the Negev Highlands prevailed for ca. 50 years, from 975 to 925 B.C.E.

Since the settlements in the Negev could not have supported themselves under the prevailing environmental conditions, they could exist only as long as they furthered the interests of the state. When the geopolitical conditions changed following the division of the United Monarchy of Israel and Shishak's campaign, settlement in the Negev Highlands ceased and was not renewed. Once the region lost its geopolitical significance, there was no justification to support settlement in the desert, and the border of Judah was pushed north, to the Beesheba Valley. This may explain the construction of Beersheba Stratum V and the fortress at Arad Level XI (A. Mazar 1985:125; Zimhoni 1985:84-7; A. Mazar and Netzer 1986:89; Ussishkin 1988:151; Yadin 1979:79-80). The Kingdom of Judah reclaimed the fortress of Tell el-Qudeirat between the eighth to sixth century B.C.E., but with very few settlements connecting it to the sedentary lands, such as the renewed settlement at the abandoned tenth century B.C.E. fortresses of Har Boqer and Har Raviv (Cohen 1981a).

Notes

1 The findings and analysis presented in this article are based on the results of the author's survey in the Negev Highlands between 1979 and 1990, conducted in the framework of the Negev Emergency Survey, and the author's MA thesis (Haiman 1988). The present article is an update of a previously published article (Haiman 1994), first presented at the 19th Annual Conference of the Department of the Land of Israel Studies in Bar-Ilan University (Haiman 1999).

REFERENCES

Aharoni, Y.
1957 *The Settlement of the Israelites in the Upper Galilee.* Jerusalem: Magness (Hebrew).
1967 Iron Age Fortresses in the Negev. *Israel Exploration Journal* 17:1-17, 97-111.
1976 The Settlement of the Tribes in the Negev - A New Picture. *Ariel* 41:3-19.

Aharoni Y., Evenari, M., Shanan, L. and Tadmor, N. H.
1960 The Ancient Desert Agriculture of the Negev. *Israel Exploration Journal* 10: 23-36.

Ahituv, S.
1979 *The Egyptian Topographical Lists Relating to the History of Palestine in the Biblical Period.* Ph.d. dissertation. Jerusalem: The Hebrew University (Hebrew with English summary).

Anati, E.
1955 *The Central Negev.* Unpublished Survey Report, Israel Antiquities Authority.

Baily, I.
1977 The Beduins in Sinai. Pp. 240-247 in *The Desert, Past-Present-Future,* ed. E. Zohar. Tel Aviv: Reshafim (Hebrew).

Banning, E. G. and Fawcett, C.
1983 Main-Land Relationships in the ancient Wadi Ziqlab. *Annual of the Department of Antiquities in Jordan* 27:375-383.

Bar-Adon, P.
1972 The Judaean Desert and Plain of Jericho. Pp. 92-149 in *Judaea, Samaria and the Golan,*

Archaeological Survey, 1967-1968, ed. M. Kochavi. Jerusalem: Carta (Hebrew).

Bartlett, J. R.
1980 *Edom and the Edomites.* Sheffield: JSOT.

Beit-Arieh,Y.
1986 Horvat Uzza - A Border Fortress in the Eastern Negev, *Qadmoniot* 73-74 :31-40 (Hebrew).

Biran A., and Cohen, R.
1981 `Aro`er in the Negev, *Eretz-Israel* 15:250-273 (Hebrew with English Summary).

Broshi, M., and Gophna, R.
1984 The Settlement and Population of Palestine in the Early Bronze II-III, *Bulletin of the American Schools of Oriental Research* 253: 41-53.

Bruins, J. H.
1985 *Desert Environment and Agriculture. in the Central Negev and Kadesh Barnea During Historical Periods.* Nijkerk: Midbar Foundation.

Cross, F. M., and Milik, J. T.
1956 Exploration in the Judaean Buqe`ah. *Bulletin of the American Schools of Oriental Research* 142:5-17.

Cohen, R.
1979 The Iron Age Fortresses in the Central Negev. *Bulletin of the American Schools of Oriental Research* 236: 63-75.
1981a The Excavations at Kadesh Barnea 1976-1978. *Biblical Archaeologist* 44: 93-107.
1981b *Archaeological Survey of Israel, Map 168.* Jerusalem: The Archaeological Survey of Israel Committee.
1985 *Archaeological Survey of Israel, Map 167.* Jerusalem: The Archaeological Survey of Israel Committee.
1986 *The Settlement of the Central Negev in Light of Archaeology and Literary Sources During the 4th-1st Millennia B.C.E..* Ph.D. Dissertation, Hebrew University, Jerusalem (Hebrew with English summary).

Dagan, Y.
In press The Judaean Shephelah Iron Age Settlement in Light of the Finds from Excavations and Archaeological Surveys. *Bulletin of the American Schools of Oriental Research.*

Dar, S.
1986 *Landscape and Pattern, An Archaeological Survey of Samaria, 800 B.C.- 636 A.D.* B.A.R. Inter. Series 308. Oxford: British Archaeological Reports.

Eitam, D.
Oil and Wine Production in Har Ephraim in the Iron Age. M.A. Thesis, Tel-Aviv University (Hebrew).
1988 The Settlement of Nomadic Tribes in the Negev Highlands During the 11th Century B.C. Pp. 313-40 in *Society and Economy in the Eastern Mediterranean (c. 1500-1000 B.C.),* eds. M. Heltzer and E. Lipinsky. Leuvan: Peeters.

Evenari, M., Aharoni, Y., Shanan, L., and Tadmor, N.H.
1958 The Ancient Desert Agriculture of the Negev, *Israel Exploration Journal* 8:241-243.

Evenari, M., Shanan, L., and Tadmor N. H.
1971 *The Negev, the Challenge of a Desert.* Cambridge, MA: Harvard University.

Faust, A.
2000 The Rural Community in Ancient Israel During Iron Age II. *Bulletin of the American Schools of Oriental Research* 317:17-39.

Felix,Y.
1963 *Agriculture in Eretz Israel in the Period of the Mishna and Talmud.* Jerusalem: Magness (Hebrew).

Finkelstein, I.
1981 Israelite and Hellenistic Farms in the Foothills and in the Yarkon Basin. *Eretz-Israel* 15:331-348 (Hebrew).
1984 The Iron Age "Fortresses" of the Negev Highlands: Sedentarization of the Nomads. *Tel-Aviv* 11:189-209.
1988 *The Archaeology of the Israelite Settlement.* Jerusalem: Israel Exploration Society.
1996 Edom in the Iron Age I. *Eretz-Israel* 23:224-229.

Finkelstein I. and Lederman, Z.
1997 *Highlands of Many Cultures.* Monograph Series of the Institute of Archaeology, No. 14. Tel Aviv: Institute of Archaeology.

Finkelstein, I. and Na`aman. N. (eds.)
1994 *From Nomadism to Monarchy.* Jerusalem: Israel Exploration Society.

Fritz, V. und Kempinski A.
1983 *Ergebnisse Der Aus der Ausgrabungen Hirbet El-Msas (Tel Masos) 1972-1975.* Wiesbaden: Harrassowitz.

Gal, Z.
1992 *The Lower Galilee in the Iron Age.* ASOR Dissertation Series No. 8. Winona Lake: Eisenbrauns.

Glueck, N.,
1959 *Rivers in the Desert.* London: Farrar, Straus and Cudahy.

Goldberg, P.
1984 Late Quarternary History of Qadesh Barnea, Northern Sinai. *Z.Geomorph., N.F.*, 28:193-217.

Gophna, R. and Singer-Avitz, L.,
1984 Iron Age I Settlement to the West of Tel Beer-Sheba. Pp. 125-131 in *Beer-Sheba II,* ed. Z. Herzog. Tel-Aviv: Tel Aviv University.

Gophna R. and Porat, Y.
1972 The Land of Ephraim and Manasseh. Pp. 196-241 in *Judaea, Samaria and the Golan, Archaeological Survey, 1967-1968,* ed. M. Kochavi. Jerusalem: Israel Exploration Society.

Gordon, R. L. Jr. and Knauf, E. A.,
1987 Er-Rumman Survey 1985. *Annual of the Department of Antiquities Jordan* 31:289-298.

Haiman, M.
1986 *Archaeological Survey of Israel, Map 198.* Jerusalem: Israel Antiquities Authority.
1988 *The Iron Age Sites of the Negev Highlands in the Light of the Negev Emergency Survey 1979-1987,* M.A. Thesis, Hebrew University. Jerusalem (Hebrew with English summary).
1989 Preliminary Report of the Western Negev Highlands Emergency Survey. *Israel Exploration Journal* 39:183-187.
1991 *Archaeological Survey of Israel, Map 200.* Jerusalem: Israel Antiquities Authority.
1993 *Archaeological Survey of Israel, Map 199.* Jerusalem: Israel Antiquities Authority.
1994 The Iron Age II Sites of the Negev Highlands. *Israel Exploration Journal* 44:36-61.
1998 *Archaeological Survey of Israel, Map 203.* Jerusalem: Israel Antiquities Authority.
1999 The Iron Age Settlement in the Negev Highlands as Reflecting the Changes in the Settlement Pattern of Iron Age I Palestine. Pp. 1-10 in *Settlement Patterns in Times of Changes,* Proceedings of the the 19th Annual Conference of the Martin (Szusz) Department of Land of Israel Studies in Bar-Ilan University, May 18th, eds. D. Hacohen and Y. Levin. Ramat Gan: Bar Ilan University (Hebrew).
In press *Survey of Kadesh-Barnea Map (470).* Jerusalem: Israel Antiquities Authority.

Herzog, Z.
1983 Enclosed Settlements in the Negeb and the Wilderness of Beer Sheva. *Bulletin of the American Schools of Oriental Research* 250:41-49.
1984 Stratigraphy and Architecture. Pp. 8-11 in *Beer Sheva II,* ed Z. Herzog . Tel Aviv: Tel-Aviv University.
1997 *Archaeology of the City.* Tel-Aviv University Monograph Series No. 13. Tel Aviv: Yass Archaeological Press.

Herzog, Z., Aharoni, M., Rainey, A. F. and Moshkovitz, S.
1984 The Israelite Fortress at Arad. *Bulletin of the American Schools of Oriental Research* 254: 1-34.

Kallai, Z.
1972 The Land of Benjamin and Mt. Ephraim. Pp. 153-193 in *Judaea, Samaria and the Golan, Archaeological Survey, 1967-1968,* ed. M. Kochavi. Jerusalem: Israel Exploration Society.

Khazanov, A. M.,
1983 *Nomads and the Outside World.* Cambridge: Cambridge University Press.

Kochavi, M.
1972 The Land of Judah. Pp.19-89 in *Judaea, Samaria and the Golan, Archaeological Survey, 1967-1968,* ed. M. Kochavi. Jerusalem: Israel Exploration Society.

Lapp, N.
1981 *The Third Campaign at Tell el-Ful: Excavations of 1964.* Annual of the American Schools of Oriental Research 45. Cambridge, MA: American Schools of Oriental Research.

Lender, Y.
1990 *Archaeological Survey of Israel, Map 196.* Jerusalem: Israel Antiquities Authority.

Maitlis, I.
1989 *Agricultural Settlement in the Vicinity of Jerusalem.* M.A. Thesis, Hebrew University, Jerusalem (Hebrew).

Malamat, A.
1983 *Israel in Biblical Times, Historical Essays.* Jerusalem: Bialik (Hebrew).

Marx, E.
1974 *Bedouin of the Negev.* Kfar Habad: Reshafim (Hebrew).

Mazar, B.
1957 The Campaign of Pharoah Shishak to Palestine. *Vetus Testamentum Supplement* 4:57-66.

Mazar, A.
1982 Iron Age Fortresses in the Judaean Hills. *Palestine Exploration Quarterly* 114:90-103.
1985 *Excavations at Tell Qasile, Part II.* Qedem 20. Jerusalem: The Hebrew University.
1994 Jerusalem and its Vicinity in the Iron I. Pp. 70-91 in *From Nomadism to Monaarchy,* eds. I. Finkelstein and N. Na`aman. Jerusalem: Israel Exploration Society.

Mazar, A., and Netzer, E.
1986 On the Israelite Fortress at Arad. *Bulletin of the American Schools of Oriental Research* 263:89.

Meshel, Z.
1974 The Negev in the Times of Judaean Kings. Ph.D. thesis, Tel-Aviv University. (Hebrew).
1979 Who Built the"Israelite Fortresses" in the Negev Highlands? *Cathedra* 11:3-44 (Hebrew).
1994 The "Aharoni Fortress" Near Quseima and the "Israelite Fortresses" in the Negev. *Bulletin of the American Schools of Oriental Research* 294:39-67.

Miller, M. J.
1990 Moab in Iron I. Pp. 131-154 in *From Nomadism to Monarchy*, eds. I. Finkelstein and N. Na'aman Jerusalem: Israel Exploration Society (Hebrew).

Moran, O., and Palmach, D.
1985 *The Water Cisterns of the Negev Highlands.* Tel Aviv: The Society for the Protection of Nature in Israel. (Hebrew).

Na'aman, N.
1988 The Southern Shefelah During the Late Bronze Age According to Cuneiform Documents. Pp. 93-98 in *Man and Environment in the Southern Shefelah, eds.* E. Stern and D. Urman. Givataim: Massada. (Hebrew).
1990 Israel, Edom and Egypt in the 10th century B.C.E. *Tel-Aviv* 19:71-93.
1998 Shishak's Campaign to the Land of Israel as Perceived by Egyptian Writings, the Bible and Archaeological Data. *Zion* 63(3):247-276.

Naroll, R.
1962 Floor Area and Settlement Population. *American Antiquity* 27: 587-589.

Negev, A.
Water in the Desert and the Nature of the Iron Age Fortresses in the Negev Highlands. *Cathedra* 11:29-37 (Hebrew).

Palumbo, G., Marby, J., and Kuijt, I.
1990 The Wadi el-Yabis Survey Report on the 1989 Field Season. *Annual of the Department of Antiquities Jordan* 34:95-118.

Rainey, A. F.
1980 The Southern Shefelah in Biblical Periods. Pp. 99-118 in *Man and Environment in the Southern Shefelah,* eds. E. Stern and D. Urman. Givataim: Massada (Hebrew).

Rosen, B.
1986 Subsistence Economy of Stratum II. Pp. 156-185 in *'Izbet Sartah,* ed. I. Finkelstein. BAR International Series 221. Oxford: British Archaeological Reports.

Rothenberg, B.
1967 *Tzfunot Negev*. Ramat Gan: Massada (Hebrew).

Shiloh,Y.
1980 The Population of Iron Age Palestine in the Light of Sample Analysis of Urban Plans, Areas and Population Density. *Bulletin of the American Schools of Oriental Research* 239: 25-35.

Shmueli, A.
1977 Beduin Settlements in the Desert Frontier. Pp. 211-216 in *The Desert, Past-Present-Future,* ed. E. Zohar. Tel Aviv: Reshafim (Hebrew).
1980 *Nomadism About to Cease.* Tel Aviv: Dvir (Hebrew).

Stager,L. E.
1976 Farming in the Judaean Desert During the Iron Age. *Bulletin of the American Schools of Oriental Research* 221:145-158.

Ussishkin, D.
1988 The Date of the Judaean Shrine at Arad. *Israel Exploration Journal* 38:151.

Woolley, C. L. and Lawrence, T. E.
1915 *The Wilderness of Zin.* Palestine Exploration Fund Annual III. London: Palestine Exploration Fund.

Worschech, U.
1990 *Die Beziehungen Moabs zu Israel und Agypten in der Eisenzeit.* Ägypten und Altes Testament 18. Wiesbaden: Harrassowitz.

Yadin, Y.
Beer-Sheva - The High-Place Destroyed by King Josiah. *Eretz-Israel* 14:79-80 (Hebrew with English summary).

Zertal, A.
1986 *Israelite Settlement in Har Manasseh.* Haifa: Haifa University (Hebrew).
1996 *The Manasseh Hill Country Survey* Vol. II. Haifa: Haifa University (Hebrew).

Zimhony, O.
1985 The Iron Age Pottery from Tel Eton. *Tel-Aviv* 12:84-87.

Zori, N.
1977 *The Land of Issachar, Archaeological Survey.* Jerusalem: Israel Exploration Society.

Zuk, Y.
1991 *Planning Farms Structures.* Tel-Aviv: Ministry of Agriculture (Hebrew with English Summary).

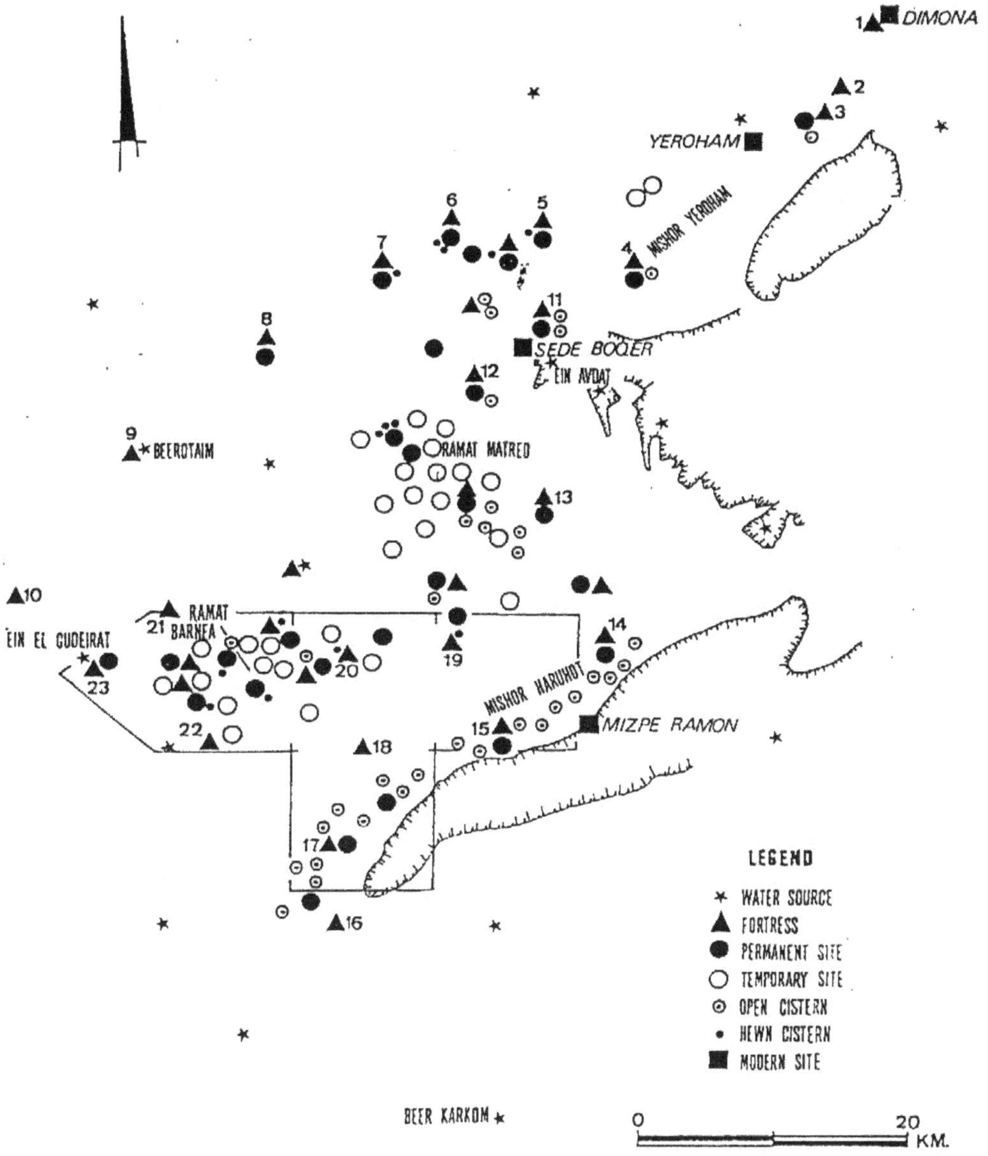

Figure 5-1: Distribution of Iron Age settlements in the Negev Highlands. The numbered fortresses are: 1) Rahba; 2) Refed; 3) Hatira; 4) Ha-Roah; 5) Nahal Boqer; 6) Mesora; 7) Shivta; 8) Har Raviv; 9) Be'erotayim; 10) Quseima; 11) Haluqim; 12) Har Arqov; 13) Nahal Zena' 14) Mishor ha-Ruhot; 15) Har Hemet; 16) Nahal Lotz; 17) Atar Ma'agorot Lotz; 18) Har Gizron; 19) Nahal Elah; 20) Nahal `Aqrav; 21) Shluhat Qadesh Barnea; 22) `Ein Qadeis; 23) Tell el-Qudeirat

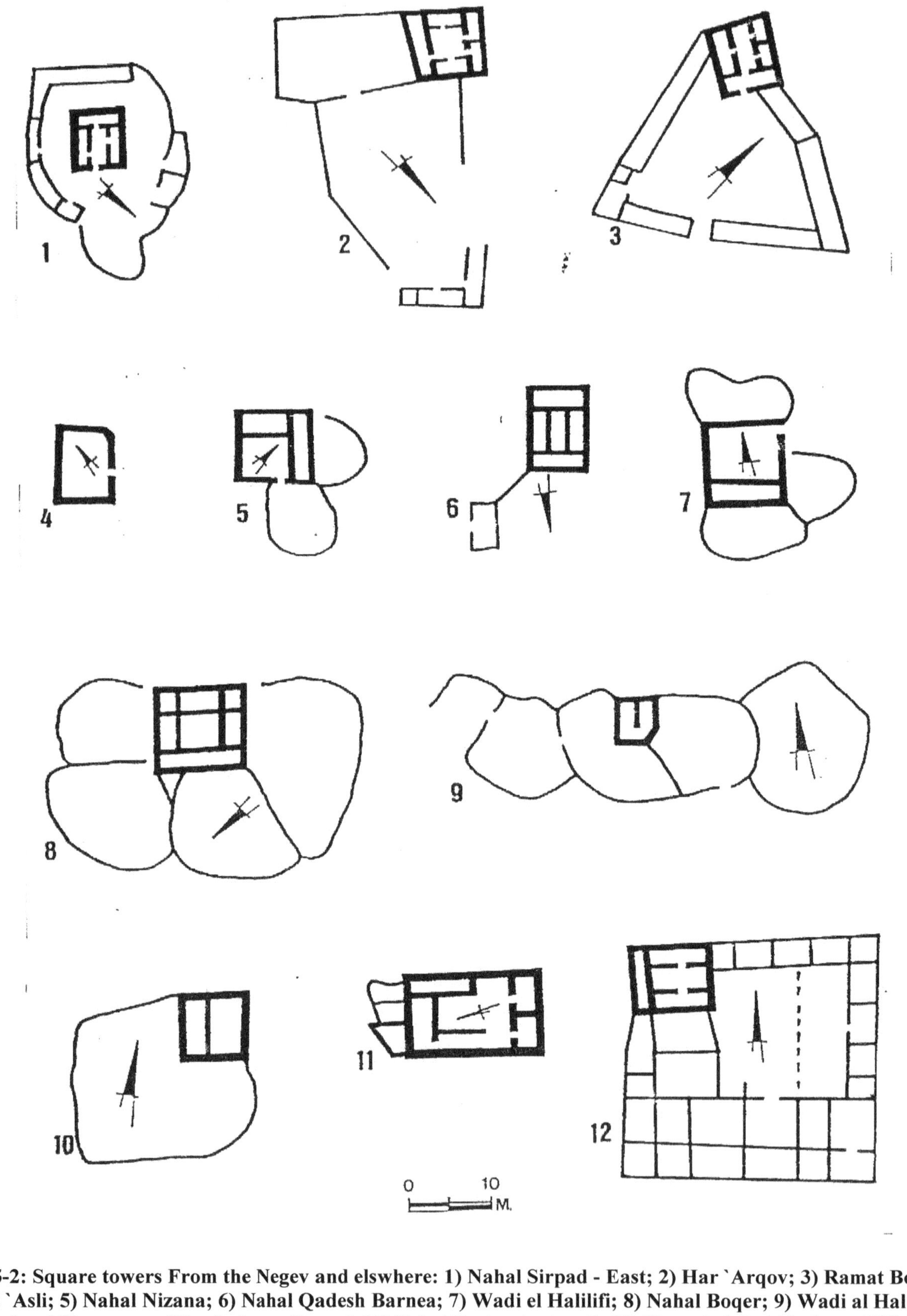

Figure 5-2: Square towers From the Negev and elswhere: 1) Nahal Sirpad - East; 2) Har `Arqov; 3) Ramat Boqer; 4) Wadi el `Asli; 5) Nahal Nizana; 6) Nahal Qadesh Barnea; 7) Wadi el Halilifi; 8) Nahal Boqer; 9) Wadi al Halufi; 10) Rujm el Hammama (The Judaean Desert) 11) `Izbet Sarta, stratum I; 12) Horvat el-Qat (Judaean Hills).

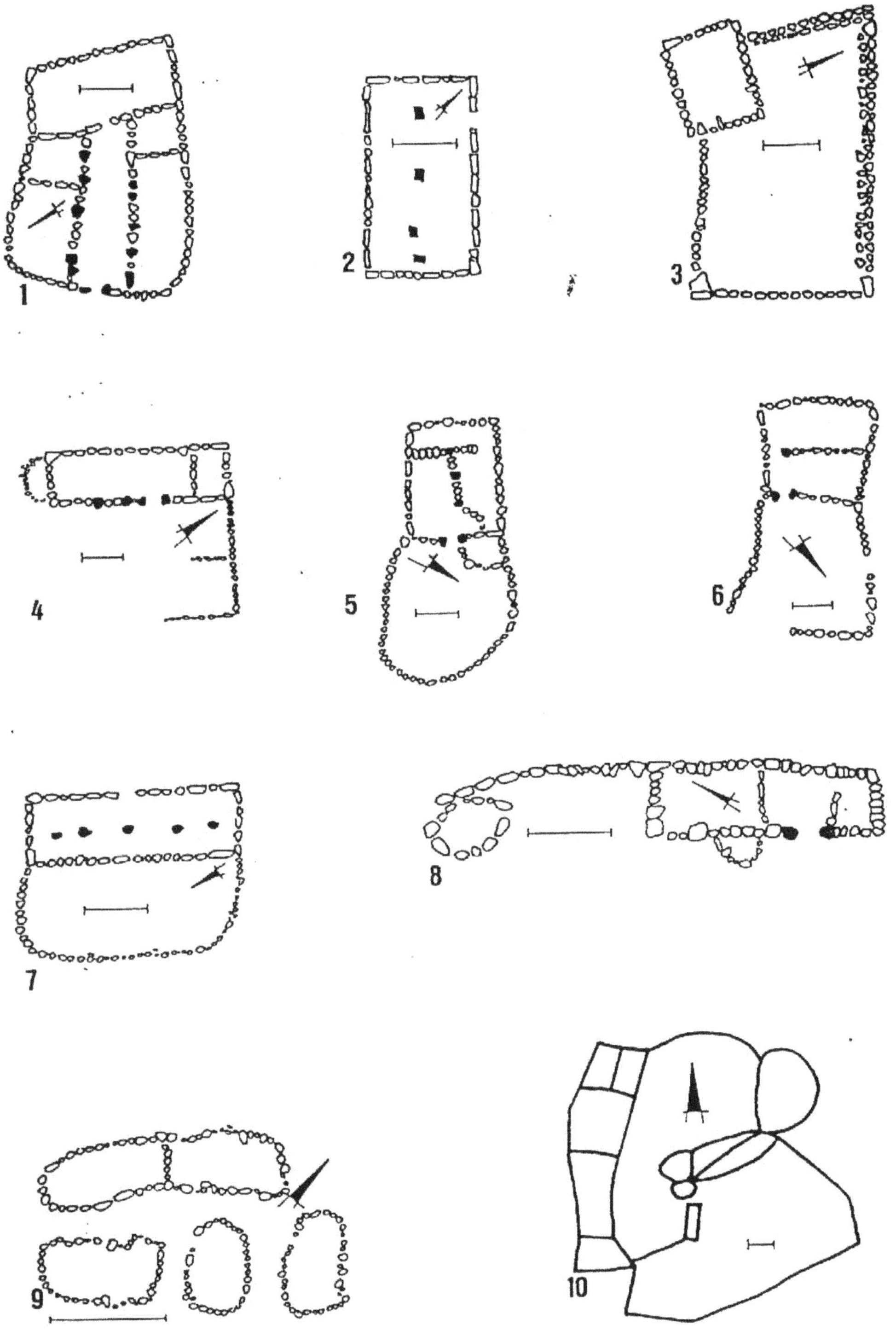

Figure 5-3: Types of structures. Group I: **1) Wadi el `Asli; 2) Nahal `Aqrav; 3) Nahal Nizana; 4) Mishor ha-Ruhot;** Group II: **5-6) Ramat Matred (After Aharoni et al 1960); 7) Nahal Elah; 8) Jebel el-Qudeirat; 9) Har Haluqim (After Cohen 1981B);** Group III: **10) Wadi el-Qudeirat; The scale for all of the buildings indicates 3 m. The dark areas indicate monolithic columns.**

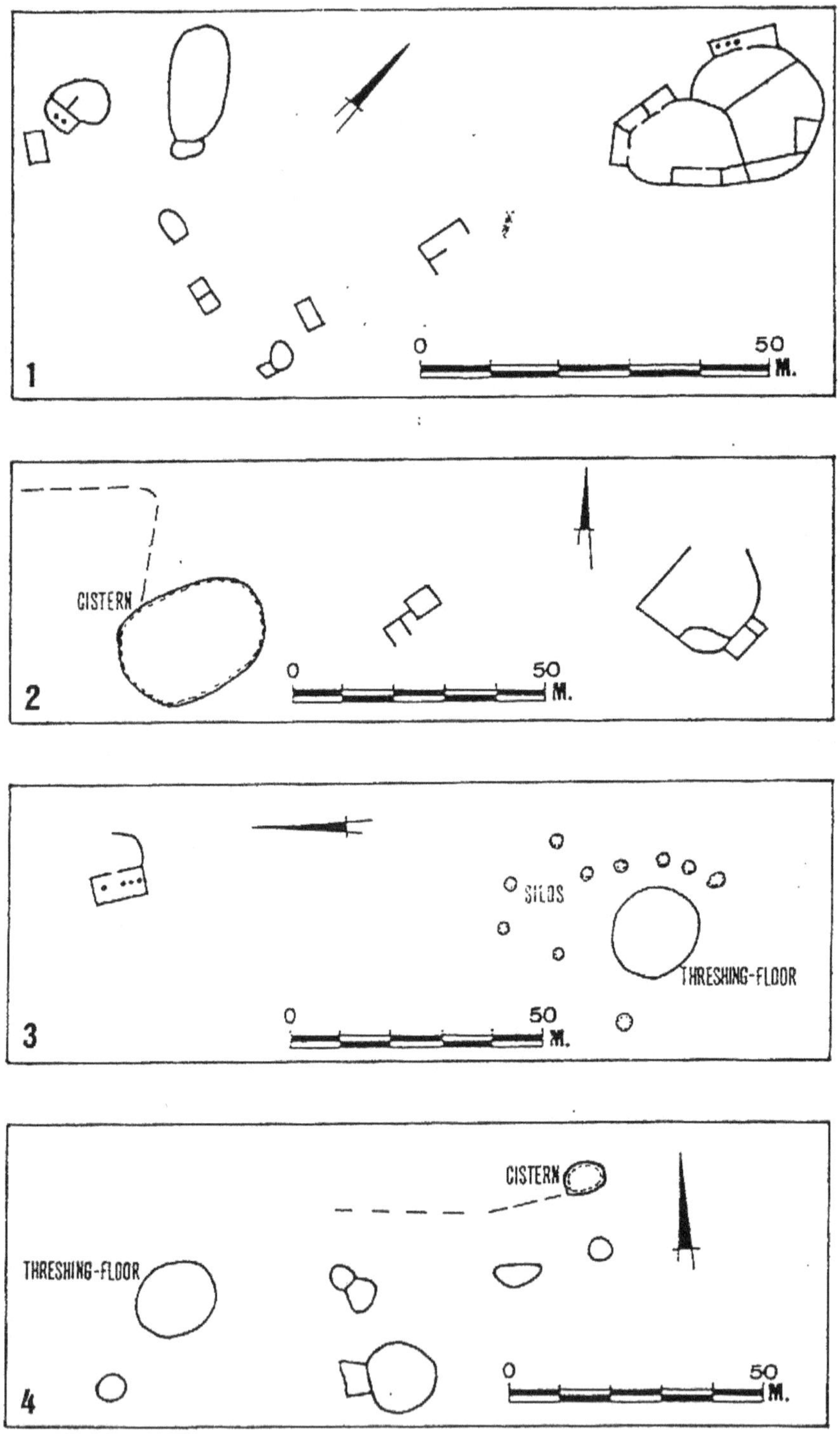

Figure 5-4: Selected settlements: 1) Nahal Elah; 2) Nahal Nizana; 3) Nahal `Aqrav; 4) Wadi um-Hashim.

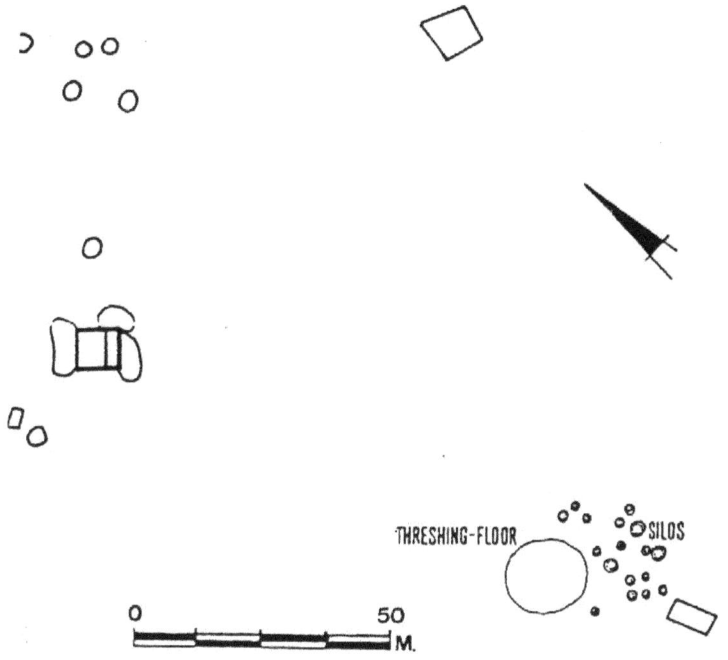

Figure 5-5: A square tower, attached by round courts, structures and installations, Wadi el-Qudeirat.

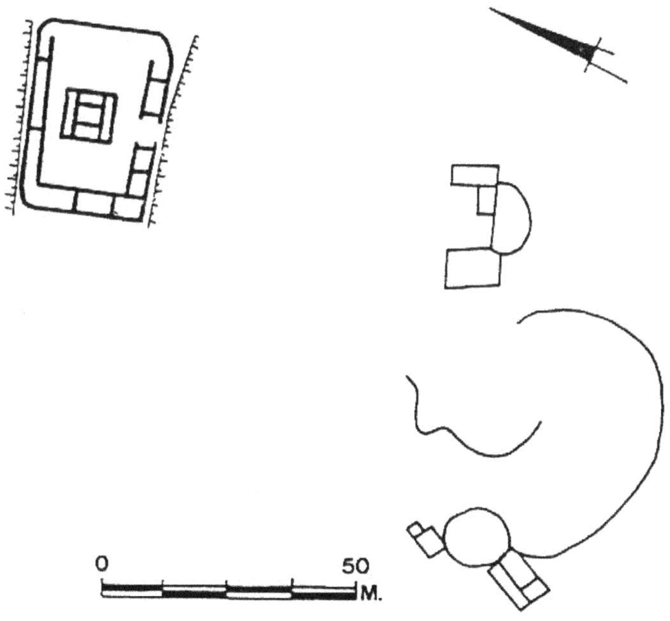

Figure 5-6: The Har Hemet fortress, with a square tower in the court and the neighboring settlement.

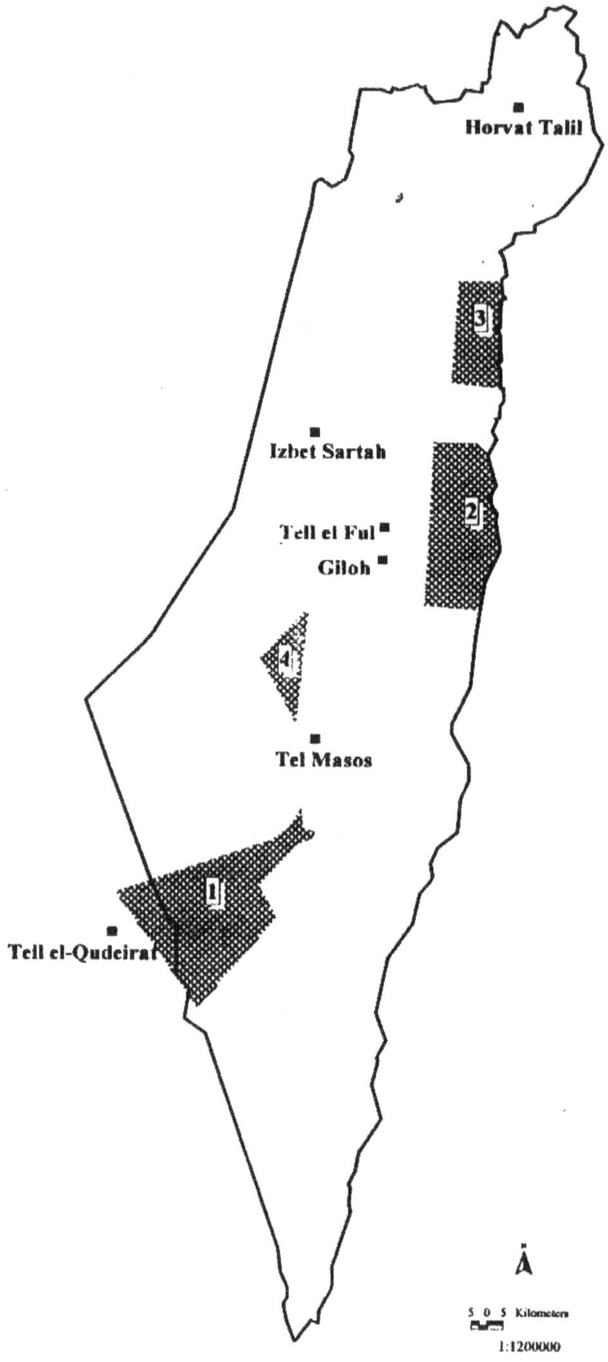

Figure 5-7: Distribution areas of the rural settlement of the type characterizing the Negev and selected sites mentioned in the article: 1. The Negev Highlands; 2. The Judaean Desert; 3. The Samarian Desert; 4. The Western Judaean Hills.

Figure 5-8: The Nahal Elah fortress.

Figure 5-9: The Nahal `Aqrav fortress.

Figure 5-10: The Har Gizron fortress.

Figure 5-11: A square tower attached by round courts viewing Wadi el-Qudeirat.

Figure 5-12: Rows of columns in a four-room-house at Wadi el-`Asli.

Figure 5-13: Open cistern at Mishor ha-Ruhot. Notice the stairs leading to the bottom.

Figure 5-14: Roofed silos at Nahal Horesha.

Figure 5-15: Open silo at Wadi el-Qudeirat.

Figure 5-16: Threshing floor combined with silos at Nahal Mitnan.

The Farmstead in the Highlands of Iron Age II Israel[1]

AVRAHAM FAUST

The Institute of Archaeology
The Martin (Szusz) Department
of Land of Israel Studies
and Archaeology
Bar Ilan University
Ramat Gan 52900
Israel
fausta@mail.biu.ac.il

This article constitutes a ***preliminary*** *attempt to summarize and analyze the data regarding Iron Age II farmsteads in the highlands. Data from various sites is studied, with special attention paid to social and economic aspects. Based on the size of structures, their internal division and the size of the agricultural area surrounding the farmsteads, it appears that the basic economic and social unit that inhabited a farmstead was usually an extended family.*
The biblical term ***haṣer*** *is also discussed, as this seems to (also) be the biblical equivalent of the modern word for 'farmstead'. In addition, regional differences in farmsteads are presented and possible explanations suggested.*

Introduction

The rural settlements of the Iron Age II, as well as of other periods, have not received much attention in the archaeological literature. This can be attributed to the "urban bias" (London 1989:41) of "tell minded" archaeology (Ahlström 1982:25). Nevertheless, considerable data from rural sites has accumulated, mainly from salvage excavations and surveys, enabling a preliminary discussion of this sector. This article will focus on one aspect of the rural sector, namely farmsteads. It should be noted that the article constitutes a preliminary attempt to summarize and analyze the phenomenon. Only with more research and more attention paid to rural sites will it be possible to confirm or refute the conclusions presented here.

Definition

According to Clark a farm is "any tract or tracts of land or of water, varying greatly in extent, worked as a unit, used for cultivation of crops or for the rearing of livestock or fish, under individual or collective management", and a farmstead is "the land and buildings of a *small* farm (my emphasis, A.F.)" (Clark 1993:114). Farmsteads, or small farms, should be differentiated from estates of various kinds. Robbers defines an estate as when "a powerful landowner has control over the people and resources of a large tract of land" (1996:57). Farmsteads are therefore isolated structures located in the midst of agricultural areas, and their inhabitants are usually the people who cultivate and use the land. Estates, in contrast, extend over large areas and belong to the wealthy and, whether the owner lives there or not, the residents may include servants, hired laborers, slaves, etc. (see also Robbers 1996:15 ff.; 56-60).

It appears that the archaeological means to differentiate between the two are the size of the agricultural area, when known, and an examination of the number, size and quality of the farm house(s), their composition and, at times, their contents.

1. An estate would have a large agricultural area, while a farmstead would have a relatively small area that could be worked by a family (small or large, see also below), the products (or income, depending on the type of economy) from which would suffice for the needs of the family unit.
2. An estate would include several buildings, or one very large building, which could house the owner and his family (whether throughout the year or only at certain times), the manager (and in many cases his family) and workers. The owner's house would be relatively opulent, and at times a difference can be seen between the residence of the owner (or manager, or both) and the dwellings of the workers. The farmstead house would suffice for a single family (and only a few workers, if any).

Farmsteads and estates represent different socio-economic systems (Safrai 1998:8-9). The estate usually belongs to a person who lives in an urban center and visits the estate several times a year. Even if the owner spent considerable time on the estate, it would remain, in this respect, an extension of the city. One of the implications of an estate is that the 'profits' leave the countryside and go to the city (for a similar, albeit not identical, system in the ancient Near East, see Magnes-Gardiner 1994). The farmstead, on the other hand, is an integral part of the rural sector. The present paper discusses only farmsteads.[2]

Farmsteads in the Iron Age II

As mentioned above, archaeological research has not paid much attention to the rural sector, and the available data on this sector in general and farmsteads in particular is relatively meager. In recent years, however, due to the construction of modern neighborhoods, settlements and roads, salvage excavations were conducted at several such sites. Additional information has accumulated from surveys, which, in some cases, can even provide sufficient data on the form, plan and size of a single-period farmstead site.

Distribution: Iron Age II farmsteads are known from most areas in the Land of Israel. Several farms were excavated in the Hebron Mountains, many were excavated in the vicinity of Jerusalem (due to the expansion of the modern city), and at least one such structure was excavated in the Benjamin region. A large group of farms was surveyed and excavated on the western slopes of the hills of Samaria.

Description of the Sites

Hebron Mountains

A farm near Wadi Fukin (map coordinates 1598\1234): A large farmstead (860 sq.m.), composed of several units, was excavated by D. Amit southeast of Wadi Fukin (Amit 1992). The eastern unit (15 x 20 m.) is probably of the four-room house type.[3] West of this structure is a wide courtyard (16 x 22 m.). Remains of walls on the edges of the courtyard probably indicate the existence of rooms. In the northwestern part of the courtyard is a square tower (9 x 9 m.). The ceramic repertoire is from the 8th-6th centuries BCE.

A farm at map coordinates 1618\1239: An almost square farmhouse (28 x 25 m.) was excavated by Amit (1992). The farmhouse is built of large field stones. The southwestern part has a courtyard, and the northern room contains a water reservoir.

el-Qatt (map ref. 1641\1187): A farm house with several large spaces and a tower attached to the northeastern corner was excavated by Amit (1989-1990). The spaces are divided into rooms in accordance with the elevations (the differences in elevation between north and south is more than 5 m.). A wine press was found near the tower. The ceramic repertoire indicates that the site was in use from the 7th-6th centuries BCE. Not far from the farmhouse, four-five additional structures, buried under heaps of stones, were discerned. According to the pottery found nearby these structures should be dated to the Iron Age.

Summary: with only three examples, it is difficult to make generalization about the form of farmsteads in this area. It is clear, however, that we are dealing with isolated structures located in an agricultural area. It appears that the structures were large and all the domestic functions were carried out in the buildings, which included a large courtyard. A tower was identified in two out of the three cases.

Jerusalem Area

Giv'at Homa (map ref. 17105\12590): The remains of two Iron Age structures were excavated by May (1999). In Area 1, a large (10 x 15 m.) structure, probably of the four-room type (three-room in this case), was found. The house had two long spaces and a square room in the back. A small wine press was found in the structure. The house was in use from the 7th-6th centuries BCE. The remains of another structure were excavated ca. 100 m. from the first, but its preservation is poor and the plan is therefore unclear.

Kh. er-Ras 1 (map ref. 1670\1282): The site was excavated by Edelstein and others (e.g., Edelstein, Gat and Gibson 1983). The excavations and surveys revealed that this is a farm complex that included a few structures, caves, pens, an agricultural area built of terraces and surrounded by a fence, and various agricultural installations dated to the late Iron Age (Kh. er-Ras 1982). The main building is a four room house (10 x 13 m.) (ibid), situated in the midst of a terraced area. A Second Temple Period wine press was found and the possibility that it was used during the Iron Age was raised (Edelstein, Gat and Gibson 1983:20-21). The site was inhabited during the 7th-6th centuries BCE (Edelstein, Gat and Gibson 1983:21). It should be noted that subsequently some of the fences were shown to be dated to the Ottoman period (Feig 1996:3). In the cave, broken Iron II storage jars were found (Maitlis 1989:60), indicating that it was probably used for storage.

Kh. er-Ras 2: About 40 m. from the first farmhouse, another four-room structure and an installation were excavated by Feig (1996). The house is quite large (more than 160 sq.m.). The broad room was probably used for storage, as it contained 300 holemouth jars. To the south of the four-room house a small olive press was identified (Feig 1996:4-6. Note that according to the excavator it is possible that the installation was originally a wine press; see Feig 1995:6), and to the north the outlines of several additional buildings were clearly visible (Feig 1996:6).[4]

Kh. 'Alona (map ref. 1677\1345): Weksler-Bdolah (1999) excavated remains from several periods at the site, which apparently included an MB village and several Iron Age II farmhouses (Weksler-Bdolah 1997:96). The Iron II remains include a three-room structure (7.5 x 12.5 m.). In front of the house are the remains of a large courtyard, with several rooms added to its south. A few installations were found in and near the structure, some of which probably pre-date the house itself. The ceramic repertoire includes late Iron Age pottery (Weksler-Bdolah 1997:97-98), but since much of it is dated to the Persian period it is likely that the house was used during this period (De Groot, personal communication), although it is not clear when the house was built. The relevancy of the data from this site to the present discussion is somewhat doubtful. The remains of several other structures were excavated, but no plans were published and the available data is very limited.

Pisgat Ze'ev A (map ref. 1730\1365): Fragmentary remains of a late Iron Age house, a wine press and a cave were excavated by Seligman (1994). "The remains and the agricultural terraces surrounding them apparently formed a farm complex, located in the rural hinterland of Jerusalem" (Seligman 1994:63). The excavator notes that "the site was severely eroded and remains consisted of poor walls and carved bedrock surfaces marking the course of the missing walls" (Seligman 1994:63). The house is composed of three long rooms, one of which was divided into two square rooms. I believe it is reasonable to suggest, given the bad preservation and in light of the plans of similar sites, that the house was a typical four-room farmhouse, with the broad room at the back not preserved. Such a reconstruction seems to correspond with the plan of the three rooms which were reconstructed by the excavator, and with the size of these spaces, as well as with other farmhouses in the area (the size of the reconstructed house would be 120-130 sq. m., similar to a few other such houses). Immediately to the north of the house, a small wine press was found, which probably sufficed only for the needs of the household. Two hundred meters to the east of the house was a cave used contemporaneously, probably for storage (Seligman 1994:63-67).[5]

French Hill (map ref. 1733\1346): The site, located east of French Hill, was surveyed by Team no. 6 of the Jerusalem Survey and excavated by Mazor and Stark (Maitlis 1989:52-57). A 1.2 ha. farmstead, surrounded by stone fences, was found. The fenced area contained agricultural terraces, an ancient road leading to the farm, cisterns and agricultural installations. There were two periods of occupation in the structure (9.5 x 11 m.), the first of which, dating to the late Iron Age, is relevant for this study. At this time, a farmhouse, large wine press, cistern and agricultural terraces were constructed (Maitlis 1989:52). The farmhouse is probably a four-room house, into which several agricultural installations were later integrated (Maitlis 1989:52).[6] A large wine press and cistern were found nearby. It is possible that part of the wine press was used for manufacturing olive oil for domestic consumption (Maitlis 1989:55-56).

Nahal Zimra (map ref. 17354\13669): The site was first excavated by Yogev, and later by Maitlis (1989; 1992). Several elements that form an agricultural complex were observed, although Maitlis notes that it is not clear whether all the elements belong to the same complex (1992:6). The complex was composed of a cistern, agricultural terraces, a wine press, a tower and a large four-room farmhouse (120 sq.m.) (Maitlis 1992:6-8). In the western part of the northern space a cave was found. It is interesting to note that ca. 50% of the rims (n=183) found in the house came from storage vessels. It is probable that the cave (which was incorporated into the house) as well as some rooms were used for storage (and it is likely that the tower was used for similar purposes).

Summary: a large number of farmsteads were excavated in this region. The typical farmstead consisted of a large four room house (average 120-130 sq. m.) ,[7] located in the midst of a terraced area. Adjacent buildings or caves were used for storage (in addition to a storage area in the farmhouse). Most farms had agricultural installations, usually wine presses, and cisterns were also found in the majority of cases.

Benjamin Region

Kh. Shilhah (map ref. 1818\1408): the site is located on the desert fringe and was discovered by Amit and Ilan during a survey, and excavated by Mazar, Amit and Ilan (1996), who reported that "The Iron Age remains are confined to a single architectural complex" (1996:194). This is a square structure with a central courtyard surrounded by at least four small chambers or rooms. The main unit, however, is a large four-room building (9.9 x 13.2 m.), incorporated within the southwestern corner of the larger structure. A cistern was found in this building. Several possibilities were raised to explain the function of the complex: a fort, a caravanserai for travelers, or the center of a private or royal estate subsisting on agriculture in the nearby riverbeds and on herding in the desert. The excavators conclude that "it is most difficult to choose between these possibilities. A combination of functions is possible" (1996:199). The inclusion of the site in this study is therefore tentative.

Samaria

Horbat 'Eli (map ref. 16410\17485): The site, located about 3 km. north-west of Shiloh, was excavated by Hizmi (1988). Several phases were identified. Two four-room houses (one of which may have been a three-room house) from the 8th-6th centuries BCE were found. One was built in the corner of a large complex that included a courtyard and several rooms along the outer walls. It appears that the two large four-room houses (the dimensions of the complete house are 12 x 14.5 m.) pre-dated the complex, the exact nature of which is unclear. A water reservoir was also found. It is interesting to note that the broad room of the complete house was probably used for storage.

Western Slopes of Samaria Hills

Many farms, dating from the late Iron Age to the Hellenistic period, were surveyed by Finkelstein (1978; 1981) and Dar (1982). In analyzing the form of the farms, Finkelstein suggests that each was an isolated, closed structure, and that although the exact plan and size of the sites varies, there are usually several similar elements: "In most farms it is possible to identify a wall that encloses a courtyard of regular or irregular shape, in which, or beside which, there is one square room or several rooms of varying dimensions... It is clear that the courtyards were not roofed...". The size of the room(s) varies from 2.5 x 2.5 m. to 5.5 x 7 m. Finkelstein adds that "In the vicinity of some of the sites, there are wine presses cut into the rock and parts of olive oil presses", and that cisterns were sometimes found in or outside the courtyards (Finkelstein 1981:335). However, as most of the farms were only surveyed, it is difficult to ascertain whether the plan remained unchanged for the entire Iron Age-Hellenistic time-span, or whether it reflects only the last period(s) of use. I will only refer to plans of surveyed structures that were in use only during the Iron Age, and not to those that continued to exist during the Persian and Hellenistic periods.

In addition, several such structures were excavated recently in salvage excavations and the combined data enables a reappraisal.[8]

Site No. 14-15 68\8 (map ref. 1468\1588): the site was surveyed by Finkelstein (1981:335, Fig. 3), who describes an isolated structure with a courtyard and several rooms, one of which contains a number of fallen monoliths. In the courtyard there is a depression, perhaps a cistern. South of the structure a wine press is cut into the rock, and nearby is a stone with a square hole, perhaps part of an oil production installation. The ceramic repertoire is from the Iron II only.

'Ofarim (map ref. 1537\1565): The site, which includes a courtyard and a structure, was excavated by Riklin (1995). The square courtyard (20.5 x 20.5 m.) contains a cistern, and in its western corner is a small room. A structure was excavated off the northern corner of the courtyard. The structure (6.5 x 9.5 m.) was probably used as a dwelling and consisted of 5 rooms. The finds are dated to the Iron Age II and Hellenistic Periods (Riklin 1994:46).

el-Bira 1 and 2: Two farmsteads, one km. apart, were excavated. The first (map ref. 1476\1589) was excavated by Scheftelowitz and Oren (1996; 1997). It includes a farmhouse dwelling (9 x 11 m.) and a courtyard (11 x 30 m.). The dwelling includes several rooms, and the excavators raise the possibility that additional rooms extended to the south (Sheftelowitz and Oren 1996:3). The courtyard is located to the north of the dwelling. Several rooms were along the western wall. The excavators suggest that the courtyard was used as a corral, and the nearby rooms for storing agricultural products. In addition, two cisterns were excavated, but it is difficult to be certain whether they were used by the Iron Age settlers or built only later. The site was founded during the 8th-7th centuries BCE, but probably continued in use through the Persian period. The second farmstead (map ref. 1565\1590) has a similar history (Oren and Scheftelowitz 1998). The farmhouse "consists of a wide courtyard (c. 168 sq m) and a residential wing (c. 133 sq m) to its south... The courtyard includes a central space... and two elongated rooms delimiting it in the west... The residential wing is composed of three rooms. The central room contained three stone bases, on which monolithic pillars were erected - apparently to buttress the roof" (Oren and Scheftelowitz 1998:50*). A simple hewn wine press was found nearby.

Tirat-Yehuda (map ref. 146\158): A very large complex (87 x 45 m.) that was established during the Iron Age and existed through the Persian and Hellenistic periods was excavated (Yeivin and Edelstein 1970). Among the Iron Age finds was a large installation for the production of olive oil (e.g., Yeivin and Edelstein 1970:58, 66; Finkelstein 1981:335; see also Eitam 1979), and it seems as if the site was a farmstead (Finkelstein 1981:335). The exact plan and nature of the Iron Age farm, however, is impossible to discern.

Summary: It appears that the farmsteads of the western slopes of Samaria were built as a single complex in the midst of the agricultural area. The complex included a dwelling\room(s), an adjacent courtyard (sometimes with additional rooms), one or more cisterns and usually also agricultural installations.[9]

Several late Iron Age farmsteads were also identified in the northwestern part of Samaria.

Kh. Zaqzuq (map ref. 1644\2097): The site was surveyed by Dar, Tepper and Safrai (1986:113-114). The farm complex is 30.5 x 30.5 m. The outer walls of the farm encircled a courtyard, partly surrounded by rooms that were used, according to the interpretation of the surveyors, as dwellings, a corral, and for storage. The ceramic repertoire is dated to the 7th-6th centuries BCE.

Moshav Reichan Farm (map ref. 1625\2085): The farm was surveyed and excavated by Dar, Tepper and Safrai (1986, 107-10). The outer walls of the complex (30.6 x 38 m.) surrounded a large courtyard, around which several rooms were excavated. It is possible that 10-15 rooms were built along the northern and southern walls, while the rest of the area was used for keeping animals and storing equipment (in an unroofed courtyard). Near the northeastern corner of the wall the foundations of a small structure were found, which the excavators suggest are the foundations of a small tower. The ceramic repertoire includes finds from the 6th-5th centuries BCE.

Summary: The farms in the northwestern part of Samaria were built as one complex, surrounded by a wall. The walled area included a large courtyard and rooms.

The Economy of the Farmsteads

The farmsteads discussed here are scattered in different regions, and it is clear that no homogeneity can be expected in terms of economic activity. Nevertheless, several observations can be offered:

Wine: It appears that wine production was the major activity of which archaeological traces were found in the Iron II farmsteads. Wine presses of various sizes -- some only for domestic purposes and others for manufacturing large quantities -- were found at el-Qatt, Giv'at Homa, Kh. er-Ras 1 (probably) and 2 (perhaps during the first phase), Pisgat Zeev A, French Hill, Nahal Zimra, Site No. 14-15 68\8, and el-Bira 2.

It should be noted that wine production was possibly the most profitable activity in the traditional economy of the Land of Israel (Safrai 1994:355-56, see below).

Olives: Olive oil production seems to have been much less common, and possible oil production installations were found only at French Hill, Kh. er-Ras 2 (later phase), Site No. 14-15 68\8, and Tirat-Yehuda.

Grazing: Grazing does not leave any clear traces. The only evidence is the presence of pens or corrals which could have been used for other purposes as well. Whether this activity played an important role in the economy of the farmsteads is somewhat more problematic; at times this can be assumed on the basis of the local ecology, e.g. at Kh. Shilhah. The excavators suggest that grazing activity also occurred at el-Bira and Zaqzuk, and pens were reported during earlier surveys in Kh. er-Ras (for this activity in the Jerusalem area, see below).

Cereals: The excavators do not mention any traces for such activity (i.e., threshing floors).

It should be noted that archaeological evidence for the last two activities are relatively difficult to find, and their absence should therefore be viewed with caution, as it is highly likely that these activities took place on the farms.

It is interesting that the major economic activity that is attested in the archaeological remains of the farmsteads is wine rather than oil production (given the presence of wine presses), as the reality reflected in the evidence from other Iron II settlements -- urban centers and villages alike -- is opposite. At these sites, olive-oil presses are found in large quantities, whereas wine presses are very rare. It appears that Isaiah 5:1-2 can be taken as an illustration of this phenomenon. In the Song of the Vineyard, Isaiah describes the activities in the vineyard:

> "My beloved had a vineyard on a very fertile hill. He digged it and cleared it of stones, and planted it with choice vines; he built a watchtower in the midst of it, and hewed out a wine vat in it...."

This song attests that all the stages of processing the grapes took place in the vineyard, and only the final product (the wine) was removed. Based on this information, there could be two explanations for the differences between the farms and other settlements (towns and villages alike): 1) people from all types of settlement processed grapes, but as the entire process took place in the vineyard, there is no evidence of this activity in settlements (as only the final product would be brought there), whereas evidence of wine presses can be found in farms, since the vineyards would be located near (or in) the farmsteads (see more on this issue below); or 2) people in settlements did not produce wine, and a kind of specialization existed with olive oil produced in settlements and wine in farmsteads.

Although the relative absence of oil production installations on the farmsteads might support the latter explanation, the former seems much more probable.[10] It should be noted that a similar situation seems to have existed in many periods (e.g. the Roman period, see Safrai 1995:206; Baruch 1998:47).

In any event, it is quite clear that most of the farmsteads participated in a market economy, and it is reasonable to assume that while they produced most of their needs, they concentrated on producing surpluses for sale/barter (in light of the above, wine probably was of major importance). Although this statement cannot be proven with regard to all sites, it is quite clear in relation to the farmsteads in the vicinity of Jerusalem. At Kh. er-Ras 2, for example, some 300 storage jars were found, attesting to the production of surpluses (see above). The excavator of Nahal Zimra notes that 50% of the rims found at the site came from storage vessels (see above).[11] The excavator of Pisgat Ze'ev mentions a cave in which 73% of the pottery assemblage was composed of holemouth jars, and it is likely that it was used for storage (Seligman 1994:71).

Another aspect of the economic subsistence of farms is discussed below.

The Social Unit of the Farmsteads

There are several means to examine the social and economic units that inhabited the farmhouse(s) and cultivated the farmland(s).

Size of the farmhouses. Many researchers have concluded that the size of a house can indicate the number of inhabitants. A considerable number of studies examining the demography of ancient Israel use the constant of 10 sq.m. (roofed area) per person (e.g. Stager 1985; following Naroll 1962; Kramer 1979; other studies, based on ethnoarchaeological data from Israel, also seem to support this figure, e.g. Finkelstein 1990).

The size of the houses on farms is usually 100-160 sq.m. It should be noted that most of the farmhouses probably had a second floor (Maitlis 1989:63-4; for a similar view regarding the four-room houses in general, not specifically in farms, see for example Stager 1985:15-17; Netzer 1992), and that there were additional structures that could have been used for housing animals and storing products. Based on the above-mentioned constant, the number of the inhabitants could have been 15-30 (Maitlis 1989:66; Dar 1986:note 21). The 10 sq.m. per person constant, however, is not universally accepted, and some use a constant of 6 sq.m. per person (e.g. Ember and Ember 1995:99). It seems reasonable to assume that the exact ratio may vary from one culture to another (see also Hayden et-al. 1996). If this assumption is correct, then it is difficult to reach a definitive conclusion as to the number of inhabitants of a structure only on the basis of a fixed constant. In the present case, however, there is an additional information that could help, as many similar houses were excavated in towns and villages. The average size of an urban house is 40-70 sq.m., approximately half the size of a farmstead house (Faust 1997a; 2000b). The fact that most of the houses are relatively small indicates that in the society under discussion, the above-mentioned ratio of 10 sq.m. per person is logical, as these houses would be inhabited by 5-6 individuals (a nuclear family). I believe this strengthens the notion that the same ratio is also applicable to the farmsteads, and the figures mentioned by Maitlis (1989:66) are realistic, i.e., these structures could house such a figure (though they did not necessarily house so many people at any given time, see below).

It should be noted that the differences in size between urban and rural houses cannot be attributed to the agricultural needs of the rural sector, as the farmsteads discussed here usually had additional spaces that could have been used to house animals and store produce. Nor can it be claimed that in towns, the houses were small due to space limitations, as small houses appear in both planned and unplanned cities, and in any event, the houses were planned to suit the needs of the inhabitants. As we are discussing farmsteads rather than estates,[12] it is not reasonable to assume that houses of more then 200 sq.m. (including second story) were used by only 5-6 individuals, especially when we know (from urban sites) that in that specific culture the required space per person was much smaller.

It is important to emphasize that the above generalization applies to all the farms under discussion, with the exception of those on the western slopes of Samaria, where some of the dwelling units are relatively small (ca. 80 sq.m.; 62 sq.m.; 99 sq.m.; 133 sq.m.; respectively. Tirat-Yehuda not included; for a much smaller estimation, based on survey only, see Finkelstein 1981:335, above). It is possible that on these farms, which could have been part of a different social and political entity, the basic social unit was a nuclear family, or that the approximate ratio of roofed area per person was somewhat different (see discussion below).

It should also be noted that not all the rural houses in the Land of Israel are so large. These farmhouses, and houses in several villages, which were probably part of a different cultural (and perhaps political and ethnic) system are somewhat smaller (for the background of the farms discussed here, see below; see also Faust 1995b; for small houses in several villages, see Faust 1995a:78-85; 2000a). This indicates that there are no economic or ecological explanations of the differences in the size of the houses in the rural and urban sectors; agricultural sites could manage with small houses. If this is correct then the explanation should be sought elsewhere, and family structure seems the most appropriate explanation (see also the extensive discussion in Faust 1997a; 2000a).

Internal division of farmhouses. the dominant house type in both the urban and the rural sectors is the four-room house, and its subtypes (perhaps with the exception of the western Samaria farmhouse, and a few sites in the northern valleys). There is, however, an important difference between the sectors in this regard, which seems to strengthen the difference mentioned above. Most of the urban houses are of the three-room house type, and are usually not divided any further. The farmsteads (like most rural houses) are usually built in the four-room style, and in most cases have additional internal divisions. The number of rooms on the ground floor only of an average rural house is as a result twice the number of rooms of an urban house, although they share the same basic plan. The larger number of rooms in rural houses should be attributed to the fact that they accommodated extended familes, and the large number of inhabitants required more segregation and privacy (especially, perhaps, between the different nuclear family units). This is strengthened by the fact that although there is *great uniformity* in the basic plan of the rural house -- a *typical* four-room house -- the internal division varies, in most cases, from one house to another. It seems reasonable to suggest that this variation is a result of the life cycle of the extended family. At some stages of this cycle, more people lived in the house and more segregation was required, while at other times, there were fewer occupants, and spaces could be enlarged again (e.g. Wilk and Rathje 1982:626; see also Seymour-Smith 1994:76; Meyers 1997:16-7).

In any event, it appears that the internal division of the farmhouses under discussion also indicates that they were inhabited by large families (again, perhaps with the exception of the farmsteads on the western slopes of Samaria).

Extent of the agricultural land. In most cases, it is impossible to identify the plot of land that belonged to a farm. In several cases, however, the excavators and surveyors were able to identify the plot and estimate its size. In the area of Jerusalem the extent of the immediate agricultural plots of two farmsteads was provided in the publications. At Kh. er-Ras 1, the agricultural area of the farmstead was identified as 0.85 ha. (according to Edelstein, Gat and Gibson 1983:21) or as 4 acres (ca. 1.6 ha., according to Edelstein and Milevski 1994:9). The agricultural area of the French Hill farm was 1.2 ha. according to Maitlis (1989:52).

Some general information was provided for the farmsteads on the western slopes of Samaria. Dar (1982) estimated the area of four of the farms at between 9.6-16 ha., and Finkelstein estimated the area of the farms he surveyed in the same area at 60.0 ha. (1981:343). It should be noted that it is not clear whether the information applies to the Iron II period or only to later periods (see above).

In any event, the area occupied by the farmsteads on the western slopes of Samaria seem to have been very large (if the information is accurate in regard to the Iron II). This can be attributed to several factors. First, it is very probable that not all the area was used and some of it was 'no-man's land'. Therefore, dividing the area among the existing farms would result in exaggerated figures. In addition, it is highly likely that grazing was among the main activities in this region, which, if so, would require large areas. Theoretically, of course, if these farms were estates, they could have extensive lands, but this does not seem probable as the structures themselves are relatively small, and lack any indication of wealth (and as already mentioned, the number of inhabitants also precludes the likelihood that these sites were estates). I therefore believe that the estimated areas quoted above are partly a result of another mode of subsistence (grazing) which seems suitable for the area, and partly due to the fact that the area was scantily inhabited and most of it was 'no-man's land' and should not be attributed to any of the farmsteads. The fact that not all of the farms existed during the Iron Age (e.g. Finkelstein 1981:346, 347) appears to support this notion, given that if the entire area were divided among the existing number of farms, their size would be enormous.

In the Jerusalem area, the size of the farmsteads mentioned above, and the high density of the farmsteads appears to confirm that these structures supported only a family unit. We do not know, however, whether the area surrounding the farm was the only area that belonged to it, or whether the farmsteads had additional plots in the nearby valleys (for fragmentations of plots see Chisholm 1979:35-6). Therefore, we cannot use these figures to determine the maximum size of the family. If there were no additional fields, it is not clear whether the areas surrounding the farms were sufficient to support an extended family. According to Safrai (1994:355), a (nuclear) family needed around eleven dumans to subsist.[13] If so, then the immediate area around the farms was insufficient for an extended family. The above figure, however, is calculated on the basis of the needs of a family in a closed (self-sufficient) economic system. The farmsteads in the Jerusalem area, however, participated in a relatively open economy, as indicated by the storage capacity of the farmsteads in general, and in particular the 300 storage jars found at Kh. er-Ras 2 (see above). Safrai (1994:355) estimates that in an open market system, a nuclear family growing *vines* can survive with *less then 0.1 ha.* Since most of these farms depended to a large extent on wine production, it appears that the immediate area would have sufficed even for an extended family. In any event, the size of the agricultural area indicates that these were farmsteads and *not estates* (even if they included other plots nearby).

At least two of the above criteria suggest that the farmhouses were inhabited by extended rather than nuclear families, and that these family units worked the land and operated the farmsteads. In addition, it appears that extended families are a more suitable unit for farmsteads, as the nuclear family is too vulnerable. The ability of the extended family to face difficulties and catastrophes is much better than that of the nuclear family (for a general discussion on extended family risk-spreading and labor optimization, see e.g. Hopkins 1985:253 ff.; see also Meyers 1997:18). According to Safrai (1998:38), the sickness or death of the father would destroy a nuclear family residing alone, while an extended family could survive because the father would be only one of the supporters. The extended family is usually also richer and has more property. Thus, it would be difficult for a nuclear family to survive *alone* on a farmstead for an extended period of time (Safrai 1998:38; note that the reality in settlements is somewhat different, as the nuclear family is expected to receive more help in such cases).

In light of the above data, it appears that extended families inhabited most of the farmsteads discussed here (with the possible exception of the farms on the western slopes of Samaria).

The Biblical *Haṣer*[14]

The word '*haṣer*' apparently has several meanings in the Bible. The first, which is completely irrelevant for our discussion, is the area around the house, i.e., a courtyard. The second meaning, according, for example, to Hamp (1986:134-35), is a small unwalled settlement. According to Malamat, the root of the word is the pre-Semitic חצר, which means "settlement," i.e. a dwelling for people and not for animals (1963:183). In many instances in the Bible and in other sources (e.g. Mari) *haṣer* is mentioned in relation to semi-nomads (Jer 49:28-33; Isa 42:1, and this concept is strengthened by the relatively wide appearance of the word as part of a settlement name in the inheritance or the tribe of Simeon [Malamat 1963:184]). According to Lowenstam "In the *haṣerim*, which are settlements that are more than nomad camps but are not walled cities yet, dwelt mainly people who subsided on grazing, such as the Ishmaelites..." (1958:273).

It seems, however, that *haşerim* existed throughout the country. On 80% of the occasions on which *haşer* appears in the Bible, it is mentioned as close to a city (e.g., in the topographical lists in the book of Joshua; Portugali 1984:283.).

While the above is undoubtedly true in most cases, it appears that the term has an additional meaning – one which did not receive a great deal of attention in past discussion:
The biblical verse that is most pertinent to this discussion is Lev 25:31:

> But the houses of the villages (*haşerim*) which have no wall around them shall be reckoned with the fields of the country; they may be redeemed, and they shall be released in the jubilee.

Many scholars have suggested, probably following this verse, that the existence of a city wall constituted a major difference between villages and towns, and that villages were unwalled (e.g. Frick 1970:44-45; Blenkinsopp 1997:54), as the *haşer* is contrasted with walled cities (as in Lev 25:29, 30). This difference seems to be supported by the frequent combination of *'ir* and its *haşerim* - the city and its villages. Some have therefore assumed that when a settlement name includes the *haşer* component (e.g., *Hazar-shu'al* in Josh 19:3), it is a daughter village of one of the cities (e.g. Frick 1970:44-5).

I believe, however, that the system described above is incorrect, probably as a result of a misunderstanding of the Biblical terms *'ir* and, (sometimes) as a consequence, *haşer*.

The word *'ir* in the Bible should be translated as 'a settlement' and not as 'a city' (or town) (e.g., Portugali 1984:284-285). This word therefore designates any concentration of houses, *including a village*. The *haşer* in Leviticus 25 is contrasted to a 'walled settlement' and not a 'walled city' as is usually translated. It should be noted that almost all the villages of the period are walled (Faust 1995a; 2000b). The *haşerim* are therefore not villages, as villages are included with the other settlements in the above verse; the *haşerim* are the farmsteads, and they stand in contrast to all settlements.

The historical reasons for the mistake are beyond the scope of this paper, but it is clear that the *haşerim* mentioned in Leviticus are farmsteads and their translation as villages (in most English translations of the Bible) is erroneous. The statement regarding the fate of the farmsteads is therefore clear: they shall be released at the jubilee because they belong to the country-field. This description accords precisely with the farmsteads presented above, *where the farmhouse is part of the farmlands*. The description does not fit the villages, as they are not part of the country-fields. Moreover, as the archaeological evidence shows, the villages were also surrounded by walls (e.g. Faust 2000b), and therefore stand in this respect with the cities in contrast to farms.

In conclusion, it appears that *haşer* means a farmstead, in addition to a courtyard. Since *haşerim* also designate, as claimed by many (see above), the initial stage in the process of nomadic settlement then the term apparently has three meanings and not only the two suggested thus. It should be stressed that the above is only a preliminary discussion. A full analysis of *haşer* and *haşerim* is beyond the scope of the present paper, and will be conducted elsewhere.

Judah and Israel: Regional Differences[15]

Most of the farmsteads discussed above are located in Judah, and only one in Israel. The situation in the western slopes of Samaria is different, and they were not part of either of these Kingdoms (see below). This difference is also demonstrated in the distribution of villages, which are abundant in Samaria, but relatively rare in Judah.

It should be born in mind that data is still relatively scarce and any final conclusions must await more information. Thus, although it appears that the almost 40 rural sites (farmsteads and villages alike), are representative, it cannot be completely ruled out that these differences are a result of insufficient data.

Safrai (1998), who studied the phenomenon for the Roman Period, described a similar situation, although on a diachronic rather than synchronic basis. According to his study, there was a decrease in the number of isolated structures over time, and he suggested several possible explanations (Safrai 1998:37-39). Not all of his explanations are applicable to this case-study, but two should be examined.

The demographic explanation. the Kingdom of Israel was much more developed and densly populated than the Kingdom of Judah (e.g., Broshi and Finkelstein 1992:53-4). This is attested by many types of evidence and accepted by biblical scholars and archaeologists alike (Miller and Hayes 1986:233-34; Oded 1984:135; Broshi and Finkelstein 1992:53-4; Finkelstein 1999). During the process of settlement growth, all available land in Israel was used, the population grew and farmsteads became villages.

The security problems explanation. Lack of security might have caused the inhabitants to prefer a settlement form of nucleated villages. However, we have no indication of security problems in the Samaria region. On the contrary, the Hebron Mountains are expected to be more problematic in this respect. Therefore, this explanation does not seem reasonable. Other explanations raised by Safrai, such as ethnic differences, are also not relevant in the present case-study (Faust 1999b; Bunimovitz and Fasut in press).

It should be noted that although Safrai discusses a diachronic difference, he raises the possibility that part of the answer might lie in regional differences (1998:39).

The demographic explanation seems the most probable. To reiterate, a definite answer, however, must await further information.

The Western Slopes of Samaria

As mentioned above, the farmsteads on the western slopes of Samaria differ in some respects from the other farms discussed in this paper.

One of the differences lies in that the dwelling units of these farms are relatively small (62-133 sq.m. for the four farms discussed here, see also the much smaller dimensions mentioned in Finkelstein 1981:335; note that the average dwelling area in the farmsteads of northwestern Samaria was probably even smaller),[16] which may indicate that the basic unit that inhabited these farms was the nuclear family, as opposed to the extended family units on most of the other farms. Another possibility is that, if the inhabitants indeed belonged to a different social group (see below), the required space per person would be less. If so, it is difficult to reach any definitive conclusion regarding the nature of the family on these farms on the basis of the size of the dwelling only.

Another difference lies in that none of the farmsteads in this area had four room-houses, while almost all the other farmhouses were built in this form. This difference stands not only in contrast to other farms, but also to other sites (including villages), some located only several kilometers away (e.g. Beit Aryeh [Riklin 1997], Deir el-Mir [Gophha and Porat 1972:232] and others), so it cannot be attributed to ecological or regional factors (Faust 1995b). Unlike the later farms in the area, however, the Iron Age farms discussed here might be interpreted as similar in some way to the four-room house, although in a *very* awkward form.[17] It should be noted that though the identification of the four-room house with the Israelites is not universally accepted, I believe its absence might be indicative of the absence of the latter (Faust 1999b; 2000a; Bunimovitz and Faust in press).

A complete discussion of these differences is beyond the scope of this article, but it appears that both could perhaps be attributed to the different background of the farmstead population. Previous studies attributed the origins of this population to refugees from the heart of Samaria after its destruction (Finkelsetin 1981), to the expansion of the highland population during the 8th century BCE (Eitam 1992:177) or to the early settlement of the Israelite tribes (Dar 1982:59). However, it appears that the source of the population lies in another, closer, area.

To begin with, I would like to refute these suggestions. It is unlikely that refugees from Samaria would come to this area. If they were to stay in Samaria, but had to leave the destroyed capital and its immediate hinterland, they would be expected to settle much closer to their original location. Moreover, even if they traveled all the way to the area under discussion here one would expect them to settle in the much better and closer area of western Samaria (only 5 km to the east), rather than on the slopes which are, ecologically, the worst area in the entire region. The western Samaria area was also destroyed or abandoned at the time (Faust 1995a, and additional bibliography there), so the refugees could have settled there. It is also unlikely that the farmstead phenomenon is part of the increase in the highland population during the 8th century, as by the end of this century the highland was destroyed, while this area flourished. If it was part of the expansion of the highland settlement, it would be expected to have shared the same fate. The fact that the opposite occurred, and when the highland settlement suffered, this area flourished, implies that it was part of a different (and even opposite) process.

It is most reasonable to assume that these farms belonged to the hinterland of the Coastal Plain, which flourished at the same time, as it did later, during the Persian period (Stern 1994:131-32; see also Finkelstein 1989:155). It is reasonable to assume that this began after the Assyrian conquest, which on the one hand, destroyed the Kingdom of Israel but on the other made this area very important as a route to Egypt (see also Porat, Dar and Applebaum 1985:58). The growth of the Coastal Plain settlement required a larger hinterland, especially when the settlements in Samaria, which probably supplied part of their agricultural needs, diminished. Although the area under discussion is ecologically poor, it is the closest area to the Coastal Plain and expansion in this direction is understandable. The historical background of the sites discussed here fits the differences between these farms and those found in the highlands.

It is possible that a combination of this scenario with one of the above-mentioned suggestions might be more accurate, i.e., that the area under discussion became part of the Coastal Plain hinterland, but some of its population originated from elsewhere in Samaria. If this is correct, it could facilitate a study of ethnic changes over time (it is clear that when the phenomenon ended in the Hellenistic Period, the settlers did not see themselves, and were not seen by others, as Israelites), via the mechanism of population transfer (albeit on a very small scale in terms of both numbers and distance) and incorporation into a different political entity in tandem with the destruction of the original political unit.

Another possibility is to attribute much of the phenomenon to a population that was exiled from Mesopotamia by Assyrians. Such a scenario accounts for the unique features of the region's farmsteads on the hand, and to some recent archaeological discoveries, attesting the presence of people from Mesopotamia in this region on the other (e.g., Na'aman and Zadok 2000). It should be noted that even if this suggestion is accepted, it is not necessarily conclusive, and it is likely that even if that was the case, a significant portion of the population was local (mainly from the Coastal Plain, but also, though probably on a smaller scale, from Samaria, see above).

The Farmsteads around Jerusalem

The farmsteads excavated around Jerusalem differ from all other in Judah, Israel and the western slopes of Samaria alike, in one respect. While these farmhouses were part of an enclosed complex, the farmhouses around Jerusalem were located in the midst of the agricultural area, and although there were other structures in the vicinity, they were not part of a single, walled, complex (see the extensive discussion in Faust 1997b).

There can be two functional reasons for surrounding a farm complex with walls:
1) Security. It appears that the main reason for building a farmhouse within a walled complex would be security problems. The walls described above could defend against thieves, encroachment, etc., but, of course, not against a real army.
2) Guarding livestock. The courtyards of the enclosed complexes could serve as corrals for livestock, although this might have been a secondary use only.

The answer as to why the farmhouses in the Jerusalem area differ from others and are not part of a walled complex appears to lie with the proximity of these farms to the urban center of Jerusalem. Jerusalem grew significantly in size and in the number of inhabitants during the 8th-7th centuries BCE. The area of the walled city reached 60.0 ha. and the entire area, including unwalled neighborhoods, probably reached 100.0 ha. (Barkay 1988; it appears that some of the unwalled neighborhoods were actually walled, see, e.g., Shukrun and Reich 1998). It seems that the hinterland became densely settled in tandem with the expansion of the urban center. Surveys have indicated the existence of about 75 sites in the vicinity of Jerusalem, most of them agricultural (e.g. Barkay 1988:125; Maitlis 1992:3), and it is reasonable to assume that the actual number was considerably larger. Some scholars have even suggested that the population that dwelt in Jerusalem and its surroundings following Sennacherib's campaign was equal to that of the rest of Judah, and this could explain the term 'Judah and Jerusalem' mentioned 16 times in biblical sources from this period (modified from Barkay 1988:125; as for the biblical sources, see, e.g., 2 Kgs 23:1; Isa 1:1; 2:1; Jer 19:7; 29:2). It appears that by the late Iron Age, the area surrounding Jerusalem was full of agricultural sites, mainly farmsteads, that constituted the city's agricultural hinterland. The situation of one farmstead bordering another, and even towns and villages, reduced the threat of thieves and robbers to a minimum (or, conversely, reflects a reality in which such a threat did not exist). The dense settlement had another important result. It forced a maximization of the use of land, and therefore intensive agriculture. This probably reduced grazing (an extensive activity) to a minimum (e.g. Efrat 1995:67-8).

It is clear from the above that the two main theoretical reasons for the existence of walled complexes -- security problems and functioning as animal pens – did not apply to the dense settlement around Jerusalem during the 8th-7th centuries BCE.

Summary

This paper is only a preliminary attempt to analyze Iron Age II farmsteads, presenting the currently available data on this form of settlement, and several basic conclusions regarding the economy and household of the farmsteads. Based mainly on the size of houses and their internal division, along with the size of agricultural land which was attributed to various farmsteads, it appears that most of the farmhouses\farmsteads housed an extended family. This conclusion is in line with studies of farmsteads in other periods. It also appears that the farmsteads discussed here were called *haṣer* in biblical Hebrew.

In addition, possible explanations for the regional differences in the form of the farmsteads are proposed:
1) Farmsteads are more common in Judah, while villages are more common in Israel. The explanation probably lies in the demographic differences between the two kingdoms -- Israel was much larger and more developed, and the same applies to its entire settlement hierarchy.
2) The farmsteads on the western slopes of Samaria differ from other farms in several respects. The differences can be attributed to the different origin and background of the settlers. These farms were part of the hinterland of the Coastal Plain, and most of the population came either from this area, or was composed from exiled from Mesopotamia.
3) The farmhouses in the Jerusalem area were not part of an enclosed complex, unlike all other farms. The explanation for this difference probably lies in the size of Jerusalem during the late Iron Age and the density of the hinterland of this 'primate city' which reduced security problems on the one hand, and lead to intensive agriculture on the other, therefore almost eliminating grazing from the local economy.

Finally, to reiterate, it should be taken into account that the data is very partial, and although several preliminary conclusions were presented, final conclusions must await further information.

Notes

1 I would like to thank Dr. Shlomo Bunimovitz, Tel Aviv University, and Profs. Ze'ev Safrai, Hannan Eshel and Shmuel Vargon, Bar-Ilan University, for their assistance and advice; the ideas expressed in this paper and any errors, however, are the author's responsibility alone. I also wish to express my gratitude to the Dr. Simon Krauthammer Chair in Archaeology, who generously sponsored this study.

2 Whether estates existed during this period is another question. It should be noted that some of the sites usually interpreted as forts could be interpreted as estates, but this discussion exceeds the scope of this paper.

3 The term 'four-room house' is used here as a generic name, and is referring to the subtypes of the four-room house as well (it should be noted that the number of rooms varies even within each of the subtypes).

4 According to the excavator, the discovery of the second house, along with the identification of several others on the same ridge, indicate that this was a village. This view, however, is somewhat problematic. The common type of village during this period is the nucleated village (Faust 2000b), and the dispersed structures do not seem to fit the typical settlement form. It is more probable that the phenomenon represents scattered farmhouses (Faust 1997b; see also below).

5 Due to the proximity of the cave also to the site of Nahal Zimra - only 400 hundred meters - the excavator was not sure whether the storage space was used by the site of Pisgat Zeev A, or that of Nahal Zimra (Seligman 1994:67).

6 Maitlis (1989:52-54) does not accept the excavators' interpretation of the structure as a four-room house, and equates it with structure 500 at Jemein (Dar 1986). It appears, however, that structure 500 is comprised of a regular pair of four room houses which is typical at Jemein. Therefore, I believe that the excavators' interpretation of the French Hill house should be accepted (at least until a final report is published).

7 The only farmstead whose area is less than 100 sq. m. is 'Alona, but, as mentioned above, it is possible that the structure in its final form is later than the Iron Age.

8 While the excavated sites were also in use from the Iron Age through the Persian or Hellenistic periods, since they were *excavated* it is possible to determine their nature during the Iron Age (unlike the surveyed sites, for which there is insufficient information to ascertain their function in each period).

9 It should be noted that a few other farmsteads were excavated in this region (e.g., Bareqet and Kh. el-Burnat 1 and 2), but either they were not published yet, or the publication is too partial to facilitate its use in the present study.

10 This is supported by the presence of wine-cellars in the village excavated at Jemein (Dar 1986).

11 Maitlis (1989:58) also mentions a relatively large number of storage jar sherds in one of the rooms at Kh. er-Ras 1.

12 On estates, which belonged to the wealthy, one could expect 'conspicuous consumption', that could result in an excessively large space for the use of the owner(s) (not all the inhabitants). This does not apply in farmhouses. Due to these structures 'simple' nature and lack of any observable luxuries, it is unlikely that they were estates. In any event, even if one claims, for whatever reason, that these structures (or some of them) were estates, he is freed from the above-mentioned constant (due to the possibility of space being consumed for 'display'), but he still refer to these houses as inhabiting a large number of individuals as an estate, by its nature, required a minimal number of workers in addition to the owner family (which was the one who could used the above-mentioned spaces). Therefore, even if these houses are to be viewed as estates (which is unlikely), then the number of inhabitants was much larger then 4-5.

13 Safrai's calculation (1994:355) is based on the relatively detailed information available from the Roman Period (from both literary and archaeological sources). There are no calculations regarding the Iron Age (or earlier periods), but although these figures relate to a later period, I believe they can be used as a general indication of the relative value of various crops.

14 I would like to stress that the following discussion is very partial. While a comprehensive discussion of the *haşer* should be conducted elsewhere, it is my aim in the present paper only to show that these farmsteads were called *haşerim*.

15 It should be stressed that this article does not attempt to describe all the farms in the Land of Israel during the Iron Age II. Some areas are not treated at all, e.g., the Shephelah, Philistia, etc. Moreover, even some structures in the Judean desert are not discussed, although it is likely that they were farms of some kind (this is because of the different nature of these farms). It should be noted that unlike most rural sites, the Judean desert's structures have been studied in detailed (e.g., Stager 1976).

16 Note that in some cases the estimated area of the western Samaria farmhouses includes the entire built area (within the complex), although parts of it were used for purposes other than living – purposes for which other farms had different structures whose area was not calculated as part of the dwelling (e.g., the towers and caves of the farms near Jerusalem). If one takes this into consideration, than the dimensional difference between the western Samaria farms and the highland farms is even greater.

17 The dwelling of the farm in site No. 14-15 68\8 (probably in the western corner of the structure) is square and not rectangular (Finkelstein 1981:figure 3), but it could be interpreted as having three long spaces divided by monoliths, and a broad room at the back. The farmhouse at 'Ofarim is a broad house rather than a long one (Riklin 1995), but could be interpreted as having an entrance into the central of the three spaces, with another space at the

back. The farmhouse at el-Bira 1 is a long house (Scheftelowitz and Oren 1996; 1997), with a broad room at the back. The long space in the front part of the structure is partially divided into long spaces with monoliths and partly with a solid wall, awkwardly forming three long spaces. The Farmhouse at Kh. el-Bira 2 is relatively large (Oren and Scheftelowitz 1998), but the western half of which could also be, in a very awkward way, interpreted as reminiscent of a four-room house.

REFERENCES

Ahlström, G.W.
1982 *Royal Administration and National Religion in Ancient Palestine.* Leiden: Brill.

Amit. D.
1989-90 El'azar. *Excavations and Surveys in Israel* 9:158-60.
1992 Farmsteads in Northern Judea (Betar Region) Survey. *Excavations and Surveys in Israel* 10:147-48.

Barkay, G.
1988 Jerusalem as a Primate City. Pp. 124-25. in *Settlements, Population and Economy in Ancient Israel*, eds. S. Bunimovitz, M. Kochavi, A. and Kasher. Tel Aviv: Tel Aviv University (Hebrew).

Baruch, E.
1998 The Economic Hinterland of Jerusalem in the Herodian Period. *Cathedra* 89:41-62 (Hebrew).

Blenkinsopp, J.
1997 The Family in First Temple Israel. Pp. 48-103, in *Families in Ancient Israel*, eds. L.G. Perdue, et al. Louisville: Westminster John Konx.

Broshi, M. and Finkelstein, I.
1992 The Population of Palestine in the Iron Age II. *Bulletin of the American Schools of Oriental Research* 287:47-60.

Bunimovitz, S. and Faust, A.,
in press Building Identity: The Four-Room House and the Israelite Mind. in *Symbiosis, Symbolism and the Power of the Past: Ancient Israel and its Neighbors from the Late Bronze Age through Roman Palestineae.* The American Schools of Oriental Research.

Chisholm, M.
1979 *Rural Settlement and Land Use.* London: Hutchinson University Library for Africa.

Clark, A.N.
1993 *The Penguin Dictionary of Geography.* Harmondsworth: Penguin Books.

Dar, S.
1982 Ancient Agricultural Farms Near Wadi Beit-'Arif. *Nofim* 16:47-60 (Hebrew).
1986 Hirbet Jemein -- a First Temple Village in Western Samaria. Pp. 13-73 in *Shomron Studies*, eds. S. Dar and Z. Safrai. Tel Aviv: Hakibbutz Hameuchad (Hebrew).

Dar, S., Tepper, Y. and Safrai, Z.
1986 *Um Rihan: A Village of the Mishnah.* Tel Aviv: Hakibbutz Hameuchad (Hebrew).

Edelstein, G. Gat, Y. and Gibson, S.
1983 Food Production and Water Storage in Jerusalem Area. *Qadmoniot* 16:16-23 (Hebrew).

Edelstein, G. and Milevski, I.
1994 The Rural Settlement of Jerusalem Re-evaluated: Surveys and Excavations in the Rephaim Valley and Mevasseret Yerushalayim. *Palestine Exploration Quarterly* 126:2-23.

Efrat, E.
1995 *Elements of Urban Geography.* Tel Aviv: Achiasaf (Hebrew).

Eitam, D.
1979 Olive *Presses* of the Israelite Period. *Tel Aviv* 6:146-55.
1992 Khirbet Khaddash - Royal Industry Village in Ancient Israel. Pp. 161-182, in *Judea and Samaria Research Studies: Proceedings of The First Annual Meeting - 1991*, eds., Z.H. Ehrlich and Y. Eshel. Jerusalem: Reuven Mas (Hebrew).

Ember, M., and Ember, C.R.
1995 Worldwide Cross-Cultural Studies and Their Relevance for Archaeology. *Journal of Archaeological Research* 3(1):87-111.

Faust, A.
1995a *The Rural Settlement in the Land of Israel during the Period of the Monarchy.* Unpublished M.A. Thesis, Bar-Ilan University, Ramat-Gan (Hebrew).
1995b Settlement on the Western Slopes of Samaria at the End of the Iron Age. Pp. 23-30 in *Judea and Samaria Research Studies, Proceedings of the Fourth Annual Meeting, 1994.* eds. Z.H. Erlich and Y. Eshel. Kedumim-Ariel: The College of Judea and Samaria (Hebrew).
1997a The Family Structure in the Villages. Pp. 131-46 in *The Village in Ancient Israel*, eds. S. Dar and Z. Safrai. Tel Aviv: Eretz (Hebrew).
1997b The Impact of Jerusalem's Expansion in the Late Iron Age on the Forms of Rural Settlements in its Vicinity. *Cathedra* 84:53-62 (Hebrew)
1999a Differences in Family Structure between Cities and Villages in the Iron Age II. *Tel Aviv* 26:233-52.

1999b *The Social Structure of the Israelite Society during the 8th-7th Centuries B.C.E. according to the Archaeological Evidence*. Unpublished Ph.D. Dissertation, Bar-Ilan University, Ramat-Gan (Hebrew).
2000a Ethnic Complexity in Northern Israel during Iron Age II, *Palestine Exploration Quarterly* 132:2-27.
2000b The Rural Community in Ancient Israel during Iron Age II, *Bulletin of the American Schools of Oriental Research* 317:17-39.

Feig, N.
1995 The Agricultural Settlement in the Jerusalem Area in Iron Age II. Pp. 3-7, in *Recent Innovations in the Study of Jerusalem*, eds. Z. Safrai and A. Faust. Ramat, Ramat Gan: Rennert Center for Jerusalem Studies.
1996 New Discoveries in the Rephaim Valley, Jerusalem. *Palestine Exploration Quarterly* 128:3-7.

Finkelstein, I.
1978 *Rural Settlement in the Foothills and the Yarkon Basin*. Unpublished Master's Thesis, Tel Aviv University, Tel Aviv (Hebrew).
1981 Israelite and Hellenistic Farms in the Foothills and in the Yarkon Basin. *Eretz-Israel* 15:331-48 (Hebrew).
1989 The Land of Ephraim Survey 1987-1989 - Preliminary Report. *Tel Aviv* 15-16:117-183.
1990 A Few Notes on Demographic Data from Recent Generations and Ethnoarchaeology. *Palestine Exploration Quarterly* 122:47-52.
1999 State Formation in Israel and Judah: A Contrast in Context, A Contrast in Trajectory. *Near Eastern Archaeology* 62 (1):35-52.

Frick, F.S.
1970 *The City in the Old Testament*. Ph.D. Dissertation (Ann Arbor, Michigan).

Gophna, R. and Porat, Y.
1972 The Land of Ephraim and Manasseh. Pp. 196-241, in *Judea, Samaria and the Golan, Archaeological Survey 1967-1968*, ed. M. Kochavi. Jerusalem: Carta (Hebrew).

Hayden, B. Reinhardt, G.A. MacDonald, R. Holmberg, D. and Crellin, D.
1996 Space per Capita and Optimal Size of Housepits. Pp. 151-64, in *People Who Lived in Big Houses: Archaeological Perspectives on Large Domestic Structures*, eds. G. Coupland and E.B. Banning. Madison: Prehistory.

Hamp, V.
1986 Haşer. Theological *Dictionary of the Old Testament* V:131-138.

Hizmi, H.
1988 Horbat 'Eli. *Excavations and Surveys in Israel* 18:51-52.

Hopkins, D.C.
1985 *The Highlands of Canaan*. The Social World of Biblical Antiquity Series 3. Bradford-on-Avon: Almond Press.

Kh. er-Ras,
1982 Kh. er-Ras. *Hadashot Arkheologiyot* 77:28-9 (Hebrew).

Kramer, C.
1979 An Archaeological View of a Contemporary Kurdish Village: Domestic Architecture, Household Size, and Wealth. Pp. 137-63, in *Ethnoarchaeology, Implication of Ethnography for Archaeology*. ed. C. Kramer. New York: Columbia University.

London, G.
1989 A Comparison of Two Contemporaneous Lifestyles of the Late Second Millenium B.C. *Bulletin of the American Schools of Oriental Research* 273:37-55.

Lowenstam, S.E.
1958 Haşer, Haşerim. *Encyclopedia Biblica* 3:273-274 (Hebrew).

Maitlis, Y.
1989 *Agricultural Settlement in the Vicinity of Jerusalem in the Late Iron Age*, Unpublished Master's Thesis, Hebrew University, Jerusalem (Hebrew).
1992 Farmsteads Agriculture Around Jerusalem in the Late First Temple Period. Pp. 3-13, in: *New Studies on the Agriculture and Economy of the Land of Israel in Antiquity*, The 12th Annual Meeting of the Department of Land of Israel Studies, ed., S. Dar. Ramat-Gan: Bar-Ilan University (Hebrew).

Magness-Gardiner, B.
1994 Urban-Rural Relations in Bronze Age Syria: Evidence from Alalah Level VII Palace Archives. Pp. 37-47 in *Archaeological Views from the Countryside: Village Communities in Early Complex Societies*, eds. G.M. Schwartz and S.E. Falconer. Washington: Smithsonian Institution.

Malamat, A.
1963 'Haşerim' in the Bible and Mari. *Yediot* 27:181-85 (Hebrew).

May, N.
1999 Giv'at Homa. *Excavations and Surveys in Israel* 19:65*-66*.

Mazar, A., Amit, D., and Ilan, Z.
1984 The "Border Road" between Michmash and Jericho, and Excavations at Hurvat Shihah. *Eretz Israel* 17:236-250.

Mazar, A., Amit, D., and Ilan, Z.
1996 Hurvat Shihah: An Iron Age Site in the Judean Desert. Pp. 193-211, in *Retrieving the Past: Essays on Archaeological Research and*

Methodology in Honor of Gus W. Beek, ed. J.D. Seger. Vinona Lake; Eisenbrauns.

Meyers, C.
1997 The Family in Early Israel. Pp. 1-47, in *Families in Ancient Israel*, eds. L.G. Perdue, et al. Louisville: Westminster John Knox Press.

Miller, J.M. and Hayes, J.H.
1986 *A History of Ancient Israel and Judah.* Philadephia: Westminster Press.

Na'aman, N. and Zadok, R.
2000 Assyrian Deportations to the Province of Samerina in Light of Two Cuneiform Tablets from Tel Hadid. *Tel Aviv* 27:159-88.

Naroll, R.
1962 Floor Area and Settlement Population. *American Antiquity* 27:587-589.

Netzer, E.
1992 Domestic Architecture in the Iron Age. Pp. 193-201 in *The Architecture of Ancient Israel from the Prehistoric to the Persian Periods*, eds. A. Kempinski and R. Reich. Jerusalem: Israel Exploration Society.

Oded, B.
1984 The Kingdoms of Israel and Judah (circa 1020-586 BCE). Pp. 99-193 in *Israel and Judah in the Biblical Period*, ed. I. Eph'al. Vol. II of *The History of Eretz-Israel*. ed. Y. Shavit. Jerusalem: Keter.

Oren, R. and Scheftelowitz, N.
1998 Khirbet el-Bira. *Excavations and Surveys in Israel* 20:50*-51*.

Porat, Y., Dar. S., and Applebaum, S.
1985 *The History and Archaeology of Emek-Hefer*, Tel Aviv: Hakibbutz Hameuchad.

Portugali, Y.
1984 'Arim, Banot, Migrashim and Haşerim: The Spatial Organization of Eretz-Israel in the 12th-10th Centuries BCE According to the Bible. *Eretz-Israel* 17:282-290 (Hebrew).

Riklin, S.
1995 'Ofarim. *Excavations and Surveys in Israel* 13:53-54.
1997 Beit Aryeh. *Atiqot* 32: 7-20 (Hebrew).

Roberts, B.K.
1996 *Landscapes of Settlement: Prehistory to Present.* London: Routledge.

Safrai, Z.
1994 *The Economy of Roman Palestine.* London: Routledge.
1995 The Shephelah of Samaria. Pp. 189-208 in *Judea and Samaria Research Studies, Proceedings of the Fourth Annual Meeting, 1994*, eds. Z.H. Erlich and Y. Eshel. Kedumim-Ariel: The College of Judea and Samaria (Hebrew).
1998 Ancient Fields Structures - The Villages in Eretz-Israel During the Roman Period. *Cathedra* 89:7-40 (Hebrew).

Scheftelovitz, N. and Oren, R.
1996 *Trial Excavations in Kh.* el-*Bira Region.* Tel-Aviv: Ramot (Hebrew).
1997 Kh. el-Bira. *Excavations and Surveys in Israel* 19:42*-3*.

Seligman, J.
1994 A Late Iron Age Farmhouse at Ras Abu Ma'aruf, Pisgat Ze'ev A. *Atiqot* 25:63-75.

Seymour-Smith, C.
1994 *Macmillan Dictionary of Anthropology.* London: Macmillan.

Shukrun, E., and Reich, R.
1998 A Wall From the End of the First Temple Period in the Eastern Part of the City of David. Pp. 14-16, in *New Studies on Jerusalem, Proceedings of the Fourth Conference*, ed. Baruch, E., Ramat-Gan: Rennert Center (Hebrew).

Stager, L.E.
1976 Farming in the Judean Desert during the Iron Age. *Bulletin of the American Schools of Oriental Research* 221:145-158.
1985 The Archaeology of the Family in Ancient Israel. *Bulletin of the American Schools of Oriental Research* 260:1-35.

Stern, E.
1994 *Dor - Ruler of the Seas.* Jerusalem: Israel Exploration Society.

Weksler-Bdolah, E.
1999 'Alona. *Excavations and Surveys in Israel* 19: 68*-70*.

Wilk R.R. and Rathje, L.R.
1982 Household Archaeology. *American Behavioral Scientist* 25:617-640.

Yeivin, Z. and Edelstein, G.
1970 Excavations at Tirat Yehuda. *Atiqot* 6:56-67 (Hebrew).

The Agrarian Structure in Palestine in the Time of the Second Temple, Mishnah, and Talmud

ZE`EV SAFRAI

The Martin (Szusz) Department
of the Land of Israel Studies
and Archaeology
Bar Ilan University
Ramat Gan, 52900
Israel

In this paper, the agrarian structure of Palestine during the Late Hellenistic, Roman and Byzantine periods is reviewed. Aspects of land ownership pertaining to private, public, royal and other types is discussed, in an attempt to define the legal and social status of various types of land throughout the region during these periods.

Land and Water Ownership

Legal vs. Social Considerations[1]

Land and water were the primary means of production in the ancient Land of Israel. An individual's economic status, and obviously his social standing as well, was dependent in great measure on the water and land resources at his disposal. Consequently, these issues, namely, of the forms of land ownership, were patently of decisive social significance.

For the purposes of the current discussion, a careful distinction should be drawn between the legislative-judicial situation, on the one hand, and, on the other, public status, or the social fabric, in which legal distinctions were interwoven. Legal nuances are of great importance, but no less meaningful is the accepted social conception and reality, although at times the latter has no legal basis. Thus, for example, the property of the king or of the kingdom was likely to be entrusted to tenants, who would become private landowners in every respect, with their formal tenancy expressed solely in a token monetary payment; also possible, however, was that in practice they would regard themselves as actual tenants, whose hold on the land was tenuous and who were at the mercy of the whim of every local official.

The research question presented in this discussion is how the agrarian structure in Jewish society in the Land of Israel under Hellenistic-Roman rule was determined: i.e. how was the mixture that fashioned the landscape defined?

Jewish law encompasses two strata: the first, the judicial, is carefully formulated in legal definitions. Jewish law, however, also contains an additional religious, intellectual-ethical stratum. For our purposes, the latter may be regarded as a sort of social conception. There is no certainty that these religious conceptions, which were preached by the Rabbis, won universal recognition and acceptance, and they may have been utopian concepts, that were not uniformly applied in actual fact, and are not to be assumed.

The complexity of the system is illustrated by the example of late Second Temple period in Jerusalem. One of the most important phenomena in the city was pilgrimage. The arrival in the city of tens of thousands of pilgrims required that accessible and available housing solutions be found. The primary accommodation solution was to stay in the courtyards of Jerusalem householders. The common practice was that, not only did the pilgrims not pay for this hospitality, but it was forbidden (or not acceptable) for householders to refuse to provide such lodgings. The hides of the sacrificed animals constituted indirect payment to the hosts, and the Rabbis even permitted the hosts to forcibly collect this payment. In other words, this was not only an accepted convention, but a quasilegal system (S. Safrai 1965:132-35).

Up to this point, such practices are of no agrarian importance. The Rabbis themselves, however, were not satisfied with this flexible definition, but rather imparted quasilegal status to the entire arrangement. According to their conception, "Jerusalem was not divided into tribal holdings" (T Maaser Sheni 1:12; PT Maaser Sheni 3:54[a], and many parallels). In other words, during the allocation of the tribal landholdings in the time of Joshua son of Nun, Jerusalem was not given to any one of the tribes, and is situated between the holdings of the tribe of Judah and the tribe of Benjamin. This fact, which is, of course, rooted in the description of the boundaries of the tribal holdings in the Book of Joshua, was given a far-reaching legal interpretation. Householders in Jerusalem presumably have no rights of land acquisition, and their possession of property in the city lacks any legal validity. Since the city is extraterritorial, the private rights of its inhabitants suffer, and they therefore are obligated to enable all the Israelites to use their property. This is patently a theoretical idea and not a legal principle, because the residents of Jerusalem were certainly entitled to purchase and sell houses in the city, and special regulations governing such transactions were enacted (M Arakhin 9:4). Accordingly, this is a religious idea that imposes on the city's residents a religious obligation with (partial but tangible) recompense. The principle that there is no private property in Jerusalem is a legal fiction that justifies the existing reality, and has no far-reaching agrarian consequences. Nonetheless, one may still examine whether this truly implies the prerogative of pilgrims to infringe on the private property rights of the city's residents. Did this right enjoy legal standing? Once again, law and social norm are intermingled here.

An even more striking example is the attitude toward the sale of a landholding. By law, a person is naturally permitted to sell his land, but, as will be shown, the sale of landholdings was unacceptable, and was frowned upon by society. Moreover, the Rabbis voiced their disapproval of the sale of family lands. The legal aspect comprised only a part, a single component, of the social order that bears examination.

For the purposes of the current discussion, the legal examination is to be separated from the social and economic inquiry (including the religious values of the Rabbis), to which we shall merely allude, and which warrants a separate study.

Land Ownership: the Legal Aspect

This essay does not purport to clarify the agrarian situation in the Biblical period, but rather seeks to examine how the law and the Biblical descriptions were likely to have influenced those living in the time of the Mishnah and Talmud. The question that we must address is: What could reasonably be understood by a Jew who read the Bible literally and sought to observe the Biblical commandments? What is required here is a simplistic, "fundamentalist," reading of what is implicit in the Bible, as reflected in the accepted interpretation in the early period itself. The second question: What survived from the early patterns, and how did those living in the period of the Mishnah and Talmud explain their deviation from what is indicated by the Bible?

The Jubilee Year - as Land Policy

The Biblical Background

The discussion should be preceded by a clarification of the status of the law of the Jubilee year,[2] which was likely to have wielded decisive influence over the agrarian structure. Inherent in the law of the Jubilee year is a revolutionary agricultural concept. A society with a law of the Jubilee year strives toward land conservatism, because all commercial movements are solely of a temporary nature. Man is the owner of his land, but he is entitled to sell it only on a temporary basis. Consequently, his ownership is limited, on the one hand, and on the other, the fear that a landless class will come into existence is also limited, because every land transaction is impermanent. The law of the Jubilee year perpetuates agrarian stratification, and probably economic stratification as well. The rich man will always remain wealthy, and is capable of straying from his class only within strict limits.

The law of the Jubilee year is apparently designed to aid primarily the weaker classes, and therefore offers a social message and an intellectual innovation. Concern for the poor recurs in the Jewish sources, and is unquestionably one of the components of this religion's uniqueness. Until the modern period, however, the treatment of poverty was localized, consisting of assistance to a specific poor person, and no attempt was made to effect a general change in society, and to fashion it in accordance with the ideal of aiding one's fellow man. Like the socialist movement, the law of the Jubilee year was a revolutionary attempt to refashion the character of society, and a compulsory moderation of social stratification.

On the other hand, the law of the Jubilee year is only suitable to a conservative society, in which economic and geographical mobility is limited. For someone who left his village and went to live in a distant city, this law was liable to be more of a hindrance than a help. This was also true for families with a large number of heirs. Under normal conditions, the inheritance is divided among the heirs, gradually creating a situation in which the landholding is too small, with some of the heirs leaving the land in return for payment. The law of the Jubilee year was likely to lead to the subdivision of the land into small plots that were not economically viable.

On several historical occasions, circumstances made a reorganization of agricultural plots possible. Such an opportunity arose when a wave of organized settlement took over a new land, or when the settlers had the ability to establish a system of laws that would express a clear ideology. This was the case during certain periods of the Greek settlement. The Greek *kolonai* instituted an egalitarian division of lands. A similar policy was adopted in the apportionment of lands in the settlement of the frontier in the United States. In both cases, this was the equality of the rich, i.e, the allocation of large plots of land to settlers of modest means (and great aspirations). In both instances, the new settlers ignored the land rights of the autochthonous settlers in the area, and in both it transpired that the initial equal distribution did not prevent the creation of economic differences in the future. The Jewish Zionist settlement also made efforts in a similar spirit, but in a different manner. The land policy of the Jewish settlement was intended to prevent private acquisition, with the goal of turning the land into national land (Kark 1989), and in this instance as well, market forces joined nonegalitarian trends to oppose the idealistic orientations.

The background against which the law of the Jubilee year emerged has not been determined. According to the Biblical testimony, this was a utopia conceived prior to settlement, as a sort of master plan that ensued from socio-religious trends in the planning stage of the settlement. Modern scholarly research, which does not accept the internal dating, generally tends to advance the Book of Leviticus to the early Second Temple period, i.e. to the beginnings of the settlement wave of the Second Temple period, or to the time of the Babylonian exile; in other words, to the phase of the utopian planning of this settlement wave. In any event, Ezekiel already mentions the "year of release" as a familiar term (Ezekiel 46:17). "Release" [*dror*] as a name for the Jubilee year appears in Leviticus 25:10, indicating that the concept of the Jubilee year preceded the settlement by those returning from the Exile.

It is noteworthy that utopian presettlement literature is a recognized literary genre that wields great influence upon historical development. *Altneuland* will serve for us as an example of the nature of the genre.

There can hardly be any doubt that the law of the Jubilee year was not practiced, and remained solely in the utopian realm. As will be shown below, it was not observed during the late Second Temple period, but some signs remain. It may have been practiced at some point in the early Second Temple period, or in the First Temple period. The law of the Jubilee year is apparently unrelated to the tribal structure, but it is connected to a society based on the extended family, and on a prevailing situation of family holdings. In the Bible it is linked to the laws of land redemption, which allude to a legal-social conception that sanctifies the family holding or, to be precise, a society that still retains social vestiges of such a concept. Consequently, an examination of the "Jubilee year" is bound up with a discussion of the concept of the family landholding.

Another law similar to that of the Jubilee year is that of the Sabbatical year. The laws of the latter mandate a cessation of working the land in the seventh year, and a declaration of the lands as ownerless for the entire year. The fruits of plantations are ownerless, and fields lie fallow. The law of the Sabbatical year is meant to restrict the right of acquisition, based on the idea that "all the earth is Mine" (Exodus 19:5), and artificially to bring about temporary economic equality. In this instance, this is not equality by the allocation of resources to all, but by the imposition of artificial poverty. The laws of the Sabbatical year were observed during most of the period of the Mishnah and Talmud (S. Safrai 1994a).

Nonetheless, the observance of the *halakhah* did not influence the agrarian structure, and it is doubtful if it changed or decisively affected the system of social stratification; these laws are therefore not included in the present discussion.

The Jubilee Year in the Reality of the Second Temple Period

According to the Talmudic *halakhah*, the Jubilee year is no longer applicable, because not all of Israel lives in its land, and the "tribes are intermingled" (*Sifrei*, *Behar* 2:2; PT Shevi'it 10:39[c]; BT Arakhin 32b; Kiddushin 38b, and parallels). This argument appears in the Talmuds, incidental to the debate concerning the status of the Sabbatical year, and to Hillel's authority to cancel de facto the remission of debts. In the Tannaitic source, however, this exposition stands by itself. Many extant testimonies attest to the actual observance of the Sabbatical year during the time of the Temple and after its destruction, but there is not a single testimony regarding the observance in practice of the commandment of the Jubilee year. It would therefore seem that the Jewish public did not comply with the latter obligation. The reason for this lack of observance is probably artificial, since the wording of Scripture does not make observance of the Jubilee year commandment dependent upon the conditions mentioned above. The Rabbis, however, understood that the commandment of the Jubilee year was suited to a clan-based social structure, and would be out of place in a relatively more developed economy based on private property.

Nonetheless, the law of the Jubilee year was not totally obliterated. The Bible connects it with the law of land redemption in a walled city. The laws of the Jubilee year do not apply in such a city, but the seller, or the redeemer (someone from the family of the seller), had the option of redeeming the land within a year. This law was observed throughout the Second Temple period, at least in Jerusalem, and was nullified after the fall of the Temple (see below).

Land Ownership in the Period of the Mishnah and the Talmud

Private Land

Private land ownership was evidently predominant in Roman Judaea. Dozens, and even hundreds, of testimonies to the purchase and sale of lands, and to the obligations of the landowner or his tenant, indicate that most of the lands in the province were owned by private individuals, both the poor and the wealthy, owners of tiny plots as well as owners of vast estates. The lands not in private ownership will be discussed below; these, however, were the exception rather than the norm in the economy of the land.

The Jewish sources provide no proof for the argument that private ownership was by its nature limited, in contrast with other legal formulations. In the legal systems of the ancient East we find the notion that the land belongs to the temple, and therefore every sale required the "approval" of the sanctuary's priests, who would check that the sale was conducted according to the law. As time passed, this examination became theoretical, and remained as a mere ceremony (Spek 1995:190).

This was also true of the ancient Roman law, in which land was perceived as belonging to the Roman urban community, and the sale of lands required the assent of the city institutions. Under pressure of circumstances, the need for approval soon turned into a mere ceremony. In practice, selling was the sole right of the seller, with the city granting its approval, as it were (Watson 1983:403-9). In each of these instances, the legal restrictions suffered a similar fate. They were transformed from a legal context to mere ceremony, and would later be canceled altogether. The desire for private property and full control of it apparently could not be curbed. The person who works the land feels that he is its owner and demands what seems to him to be his natural right to control the land and to derive benefit from it without limitation. At any rate, in the period under discussion, land was likely to be completely within the private domain under Roman law as well. Moreover, the surveying of land and the determination of landholding boundaries was regarded in neighboring Christian Syria as an act conducted on behalf of God (Voeoebus 1977:153-54; 1982). Interestingly, in the Syrian-Roman law book the division was perceived as a civil activity of the emperor, and in a church law book it is presented as an act of God. Thus private property wins the blessing of God, thereby imparting new content to the concept of "the earth is the Lord's and all that it holds" (Psalms 24:1). "For all the earth is Mine" (Exodus 19:5) - and by My will the land is awarded to everyone, both to the person who inherited and to the one who purchased.

One would expect Jewish thought to develop the idea that the land belongs to God, with various Biblical verses providing the foundation for such an idea. An expression such as "For all the earth is Mine" (see above) might be understood as conveying this concept. The same holds true for statements by the prophets, such as Hosea, who declares: "Assuredly, I will take back My new grain in its time, and My new wine in its season" (Hosea 2:11), implying that the new grain and the new wine belong to God. The Rabbinic literature makes use of this idea to explain man's obligation to pay tithes. It is as if man does not give one tenth to the Master of the Universe, but rather receives eight or nine tenths from the Lord as a personal gift from Him (*Pesikta de-Rav Kahana, Aser Te'aser* 1:162, and parallels). All this, however, remained solely on the abstract and theoretical plane. In practice, ownership was unlimited, with the exception of ecological restrictions, or those ensuing from the laws of damages.

Later, in the Mamluk period, or possibly only in the early Islamic period, this concept would return (Frankel 1996), and it is prominent in Islamic law and up to the present. This perception may have been influenced by a prevailing situation of communally owned tribal lands (see below). According to this concept, ownership by the individual is limited, and he is no more than a person with certain rights to the land. This then led to the formulation of the notion that the individual could not acquire the primary means of production at man's disposal. Needless to say, these socio-legal concepts were unsuccessful in inhibiting the creation of private property and of economic differences within society.

In Egypt, which has yielded much legal material, there was a secondary classification of the types of private land, based on legal standing; in other words, according to the manner in which it came into the owner's possession (inheritance from one's parents, a gift from the ruler, etc.), and according to the uses of such land. Each of these categories was expressed in a different rate of taxation. In addition, we find different names for land in accordance with the amount of tax collected from it, even without reference to the preceding distinctions (Rowlandson 1996:27-69, and particularly 41-55).

In the Land of Israel all private land was included in a single legal category. Or, at the very least, there is no extant allusion to a legal classification of private holdings, as was the case in Syria as well. This may be a consequence of the paucity of extant material from these provinces, but it would seem that we nevertheless possess sufficient information, and we may conclude that the situation in Egypt was different. Generally speaking, the Egyptian bureaucracy was especially well-developed and cumbersome, and it would seem that no conclusions regarding the other parts of the empire may be automatically drawn from the land of the Nile.

Family Lands

The landholding system in the First Temple period was based on the rights of the extended family to lands, along with a parallel minimal land mobility. These two elements are interconnected and mutually consequential, but not identical. On the one hand, family lands could be sold, while, on the other, the restrictions could also be imposed on the sale of private land. The constraints on land sales during and after the Second Temple period will be discussed below; the question to be addressed now is whether traces of the family ownership of the land remained.

Society underwent changes in the late Second Temple period, and after a lengthy process, the extended family was reduced, and the nuclear family became increasingly prevalent (Z. Safrai 1983; Rubin 1972). We should therefore not expect to find in the Tannaitic period a land system based on family holdings. Indeed, in a majority of the sources, land is perceived as private property, with only moral and emotional restrictions applying to its sale. Nonetheless, a number of allusions indicate that in this period as well, there are a number of traces of a different social reality.

A *baraita* in Tractate *Semahot* determines: "All threshing floors that are inherited may be moved from place to place and transferred from one family to another. A tomb may be neither moved from place to place nor transferred from one family to another" (*Semahot* 14:2, [ed. Higger]: 204). The tomb is therefore a family possession, and may not be sold. The private householder is not its owner, since it belongs to his family for all time. On the other hand, a family threshing floor may be uprooted and moved, and is not considered family property. The very term "threshing floors that are inherited," the emphasis upon "the uprooting of the family," and the need for permission to sell the threshing floor, attest that family property was still of significance, and that full ownership by the individual was not yet self-understood. This *halakhah* is characteristic of a period in which the family had already ceased to function as landowner, but was still an element to be reckoned with - less on legal than on social grounds. It was only natural that the former rules would still apply to the tomb.

A family tomb is mentioned in another *halakhah*, which appears three times: "If a [person] has sold his [family] tomb, the path to the tomb, his halting place [for consolation], or his house of mourning, the members of his family come and bury him by force, because of the slight against the family" (BT Ketubot 84a; Bava Batra 100b; Bekhorot 52b). The tomb was accordingly regarded as the property of the family, with the latter being considered an economic unit, albeit without a clear and decisive definition. This *halakhah* appears only in the Babylonian Talmud in the name of Tannaim, but it would seem to represent the Land of Israel as well, since the concept is also found in *Semahot*, which includes mostly material from the Land of Israel, and contains many early traditions.

Additionally: "From where is it derived that the one who sells the tomb of his forefathers transgresses a positive commandment? Scripture teaches ..." (*Sifrei*, Deuteronomy 188, p. 227; *Midrash Tannaim* on Deuteronomy 19:14, p. 115). A hint of the concept of family property may also be discerned in another halakhah: "If one inherits his wife's estate, he returns it to the members of the family, and gives them a deduction from the purchase price" (M Bekhorot 8:9), or: "The members of the family give him money and take it from him" (T Bekhorot 6:19; BT Bekhorot 52b). The Talmudic discussion is related to the theoretical context of the Jubilee year, but between the lines we seem to hear an echo of the convention that family property was of significance in the time of the Tannaim as well. The family as an economic body appears in many other *halakhot*. Almost all the testimonies refer to the extended family, that is, to "brothers who are partners" (Z. Safrai 1983), but at times mention is also made of a broader and less well-defined family. Thus, for example, we hear of a "family flaw" ensuing from one of its members having a defect. Thus, the Palestinian Talmud explains a mishnah (M Bava Kamma 8:7; PT Bava Kamma 8:6[c]), and concludes that "shame is given to the relatives," that is to say, the members of the family are entitled to compensation for this shame. Once again, these are not brothers, but "relatives" in a considerably more general sense. Incidentally, the subject of a "family flaw" became much more central and important in Babylonia, due to the special perception in the Diaspora of the importance of family lineage, but this is not pertinent to the current discussion.

Consequently, we can conclude that in general, private property existed, or belonged to a limited extended family. Vestiges still remained, however, of a different legal context, of the property belonging to the large and undefined clan. In the past, during the early Hellenistic period, or even earlier, this system of land ownership was much more common, and the "extended family," or even the "clan," constituted the ancient societal context, although there are few extant proofs of this.

Public Lands

Communal Lands

Prevalent in the land system of the East during the medieval period was the special *musha'a* ownership system. In this system, lands belonged to the members of the village as a whole. Each year a lottery was conducted, with the lands being allotted among the members of the village or tribe according to the results of the lottery. Individuals did not have the right of ownership, but only that of usage (Grossman 1996). There is no allusion to such an agrarian structure in the Jewish tradition, nor are there clear testimonies throughout the roman East. One inscription from eastern Syria mentions the leasing of lands. The lessees are the members of a single tribe, who leased the land or the estate (*epoikon*) from one of the rural communities (IGR 3:132; for the term *epoikon*, see the extensive discussion, below). MacAdam, who examined this, saw in this inscription an allusion to communal ownership by the villagers, and, more importantly, joint utilization of the area (MacAdam 1984). This is still not *musha'a*, but rather common ownership and working of the lands. This apparently refers to wilderness pasture land, which in his opinion, is an important element in understanding the regional agrarian system. In another inscription, the villagers agree that no member of the community will graze on the common land. The use to be made of the land is not specified in the document, but, once again, the land is communal, and is not given over to individuals to be worked (Waddington 1870: No. 2505).

The joint ownership by the community of its lands is connected to the socio-legal system in which the community or city possesses its lands. Since these are public lands, the public is entitled to divide them among its citizens, or to continue to hold them as a communal possession, and to establish rules for their temporary allocation among those who will work the land, or to lease them to whoever offers the highest rental. Such lands were a common phenomenon under Roman rule (see below). In the Hellenistic and Roman-Byzantine system, public property was administered in a completely different manner. Nonetheless, it is understandable how communal ownership evolved into the joint working of the land.

A small number of Egyptian papyri may be describing the allocation of the rights to work the lands belonging to the village. This allocation was made for a limited time, and was determined by means of a lottery, known as *pitki*. Also mentioned is the parallel joint working of public lands, under the administration of the village elders (Rowlandson 1996: 80-82). This method is quite rare, but is well documented. The allocation of the landholdings in the time of Joshua is depicted in a midrash as the "classic" land allocation of the Greek *kolonia* (BT Bava Batra 122a; PT Yoma 2:41[b]). It cannot be inferred from this, however, that the method of joint working of lands was practiced in the Land of Israel. Here, too, the allocation of was implemented by means of a lottery, as is written explicitly in the Bible, but this was a fixed apportioning, and not one that changed annually. It formed the basis for "eternal" private ownership, and not for the right of one-year use.

In contrast with the view of MacAdam, it appears that the testimonies concerning the joint working of land are few, and that the distribution of this method was quite confined. The testimonies from Egypt also are very limited. Moreover, the absence of evidence regarding this method in the Land of Israel allows us to assume that this practice was not common there, and certainly was not prevalent.

The allocation of lands on a rotation basis is suitable for the utilization of grazing land, or at best, for the intensive working of sown grains. It is not incidental that the two testimonies cited by MacAdam (and primarily the first testimony, which is more certain) relate to the working of wilderness land, which is suitable mainly for grazing. The changing ownership serves to hamper investments in the land. Who would invest his energy in land that will be given to another farmer the following year? In hill country land, which requires the constant removal of stones and the construction of terraces, changing ownership leads directly to deterioration. Whatever the reason, there is no hint that the method existed or was widespread in the Land of Israel.

Royal Estates

When the Romans took formal control of the Land of Israel in 6 CE, they found a ramified system of royal estates. Royal lands were already a prevalent phenomenon in the Biblical and Persian periods. The Persian crown estates passed directly into the hands of the Ptolemaic and Seleucid governments. During the Hasmonean revolt, there was of a tax in the amount of 50% of the orchard crop, and one third of the yield from the grain fields (I Maccabees 10:30). Such a tax is considerably higher than that imposed on private lands, but corresponds to that imposed on the crown estates in Egypt and in Syria. Scholars have accordingly surmised that all the lands of Judea became imperial estates; and apparently in the wake of the revolt, all the lands in the rebellious land were expropriated (Bickerman 1979:9-17; Mittwoch 1955).

If all the lands of Judea were in fact confiscated, the success of the uprising turned this seizure into a fiction. Nonetheless, the Hasmonean rulers were likely to have transferred to their possession the already existing crown estates in Judea. An example of this is the estate in Jericho. A government estate, which included the Balsam (*afarsimon)* areas, undoubtedly continued to be active in the Jericho region. In this region, the Hasmonean kings later built a "palace" that included agricultural installations, and was merely a magnificent estate house of the royal owner (for a summary, see: Z. Safrai 1994: 147-55).

The plethora of wars and conquests in the period 110-76 BCE provided a wealth of opportunities for enlarging the crown properties. The areas that were conquered and the Greek cities whose inhabitants were expelled, were likely to have become royal lands. There are no extant direct testimonies concerning the degree to which the rulers of the Hasmonean dynasty exploited these possibilities, but such actions are highly probable. Herod, the successor of the rulers from the Hasmonean line, granted his soldiers lands in Gaba of the Cavalry, and in Sebastea (Schalit 1969:184-89), from what were probably royal lands. An additional royal estate was situated in Iamnia (see below), which also attests to the existence of crown possessions in the vicinity of these cities.

The arrival of the Romans in 63 BCE did not alter the situation in Judea itself, but changes probably took place in the lands within the bounds of the cities that were restored and rebuilt, changes that enabled the settlement of new elements in these cities.

Royal lands were a prevalent and accepted phenomenon in the Roman Empire. In legal terms, there were a number of types of such lands: those that belonged to the empire, those that were the personal possession of the emperor, and possibly additional categories as well. The terms also lost their original meaning. Thus, for example, the term *ousia* was first used to describe the estate of the emperor, but became an appellation for any private estate (Rowlandson 1996:55-57). None of these distinctions, despite their legal importance, were of social or organizational-economic import. In Egypt, with its lengthy tradition of ownership of land by the monarch, the royal estates constituted an extremely important economic element. It is difficult to determine what percentage of the land was included in this category. Based on a number of documents attesting to land registration in certain villages, one scholar has estimated that about 80 percent of the land in Egypt belonged to the king (Rowlandson 1996:63). This estimate may be exaggerated, because naturally, fewer documents of private landowners who worked their own lands would have been preserved. At any rate, it is clear that the royal estates comprised an important segment of all the lands in the province. There are many documents pertaining to the organization of such estates, but there is no certainty that the same system was employed in Judea (Parassoglou 1978). Much information may be drawn from a number of inscriptions that attest to royal estates in Africa (Kehoe 1992). In general terms, the royal estates were managed in the same manner as the private estates; this will be reexamined in the discussion of the latter.

In the Roman and the Byzantine empires, all quarries and special natural resources were considered the property of the emperor: silver, copper, and gold mines, and all the marble and salt quarries. The Lebanon forest was a royal estate, at least from the time of Hadrian. It may be assumed that common stone quarries were privately owned, while the salt mines in the Dead Sea region and the evaporation pools along the Mediterranean Sea were probably royal property.

An interesting phenomenon can be discerned in Egypt: Royal estates were leased at relatively low rates (Rowlandson 1996: 75-78). The average sum was an *artaba* (about 40 liters) of wheat per *aurora* (2.7 dunams), while private land was leased for 1.5 *artaba* per *aurora*, or even more. This testimony joins other allusions, to be discussed below, concerning the less efficient management of public property.

There are many historical and epigraphic testimonies from the nearby provinces regarding royal lands. We will mention only a single inscription from Syria that delineates the boundary between the *timion*, a common term for royal property, and a certain private estate (Prentice 1908: No. 28a).

In this period, i.e. the late Second Temple period, there are reports of extensive territories that were royal estates in Judea. Herod possessed extended lands (Schalit 1969:135-36, and more) and Agrippa, his grandson, owned estates in the vicinity of the Jezreel Valley. The center of this estate was in Beth Shearim (Josephus, *Life* 24), and even earlier, Hyrcanus II may have possessed an estate in the Jezreel Valley (*Ant.* 14:207; Mazar 1940:15). An additional estate was most probably located in Shihin, although it is unclear whether it belonged to Agrippa II, or to the emperor himself (T Shabbat 13[14]:9; PT Shabbat 9:15[d]; BT Shabbat 121a). During the War of Destruction (66 CE) Giscala was a center for produce taxes, but this does not constitute proof of the existence of a royal estate; it may have been merely a toparchy capital (Josephus, *Life* 13). Herod gave Iamnia to his sister Salome, who transferred it to Livia (*Ant.* 17:189, 321; *War* 2:98; *Ant.* 18:31; *War* 2:167). Somewhat later, mention is made of the imperial *epitropos* in Iamnia, who took control of the estate and administered it (*Ant.* 18:150; cf. Philo, *The Embassy to Gauis* 199). A tomb inscription of the daughter of an imperial procurator was discovered in this city (Avi-Yonah 1946:84). In the fifth century, the literature again speaks of the *epitropos* of the king (*Petrus the Iberian* 123), and mention is also made in this context of a hostel built by the empress Eudocia in the city. Nevertheless, not all of Iamnia was a royal estate, since the place appears as a *polis* already in the first century (Avi-Yonah 1963:94), and the Jewish sources mention individuals who were private landowners in the city, such as Rabban Gamaliel and other rabbis (e.g., M Demai 3:1). The concept "mountain of the king" appears in the sources, and is indicative of royal lands (Applebaum 1967). It should not be inferred that the entire area was a royal estate, although such estates were prevalent in the region.

Josephus relates that after the revolt, Vespasian expropriated all the lands of the province, which presumably leads us to conclude that the entire province became a crown estate (*War* 7:216). S. Safrai has already demonstrated, however, that in practice not all the lands were the property of the emperor, and in the generation following the uprising there is a proliferation of reports relating to private lands in Judea (S. Safrai 1994:306-10; Buechler 1912:113-30). The legal structure dictating that the lands of a rebellious province belonged to the imperial treasury is to be understood as a legal theory and as a political option, and not as a realistic rule that was actually implemented in Judea. Isaac continued this reasoning (Isaac 1984) and, based on an analysis of Josephus' descriptions, determined that not even he intended to claim that all the land of the province had been seized. At any rate, as will be shown below, private landowners undoubtedly were still present in the province in the post-revolt generation, and many sources mention the purchase, sale, or rental of land.

The Bar Kokhba revolt, which was of a much greater intensity than the War of Destruction, provided an additional opportunity to increase the royal lands. Eusebius expressly states that Rufinus expropriated all the lands of the uprising (*HE* 4:61). There are also reports from this period of a large vineyard that belonged to Hadrian (PT Taanit 4:69[a]; Lamentations Rabbah 2:2). In this time, either before or after the Bar Kokhba rebellion, the *metzikim* [oppressors] appear in the Talmudic sources. Applebaum identifies them with the *conductors*, the officials in charge of the immovable property of the emperor. In this period a new *halakhic* term was born: the law of the *sikarikon*, that relates, in the view of Safrai, to the lands expropriated from private farmers on behalf of the imperial treasury (S. Safrai 1994:159-167).

Many lands were undoubtedly confiscated in the wake of the rebellion. The property of prominent rebels and those known to be wealthy was seized, just as the possessions of Archelaus had been expropriated in the past (6 CE - *War* 2:111; *Ant.* 17:344; 18:26). In this post-revolt period there are reports of many tenants and farmers who were stripped of their property. Beyond our general knowledge, however, there is no clear information that would indicate the extent of the crown's landholdings.

Such royal estates were probably seized by Bar Kokhba during his uprising, and some of the documents of leasing agreements were discovered in the Judean Desert caves (Milik et al. 1961: No. 24; Applebaum 1967).

It was common practice in the Roman Empire to grant lands to veteran soldiers from the stock of landholdings in the crown estates. However, hardly any lands in Palestine were allocated to veteran soldiers. We know of colonies of veterans that were established near Jerusalem following the War of Destruction (*War* 7:216), of a single veteran in Azzun in Samaria (Bull 1966), and of a veteran, or colony of veterans, in Aphek in the Golan (Mann 1983:43 ff.). There is no extant information concerning additional veterans. Other lands were awarded to local wealthy individuals, such as Rabbi Judah ha-Nasi, who received lands in the Golan from the emperor (PT Shevi'it 6:36[d]). We also know the term *kerem doron* (literally, a gift vineyard; PT Peah 7:20[a]), that may possibly indicate a vineyard that was given to a local rich man. Once again, there is no extant information concerning the extent of these grants.

Another question, of greater political interest, is what became of the "private-royal" property of Herod. This was given to his heirs, but what happened following the War of Destruction?

The information from the Roman-Byzantine period is not plentiful. Byzantine sources mention a number of villages that belong in their entirety to the emperor, such as Kfar Ginata (Petrus the Iberian 1895: 20-39). Another village, Porphirion, a private village belonging to one of the wealthy inhabitants of Caesaria, was confiscated by the emperor Justinian, with no legal justification (Procopius, *Anecdota* 30:18-19; 24-30). Such arbitrary seizures were the instrument for increasing crown possessions, despite the protests by the inhabitants of the province. The village of Turban also was most likely royal property, which was administered in practice by a chief custodian (Zacharias Rhetor, 1899: 262 and parallels); such an official, an indigenous Samaritan, is mentioned in the writings of Procopius (*Anecdota* 27:31). A Latin inscription mentioning the procurator was found in Bir el-Abad. The researchers surmised that this attests to an imperial estate, but this is no more than conjecture (Applebaum et al. 1978:137-38).

The administrative lists of Palestine include two regions termed *saltus*, meaning an extended imperial estate that was an independent administrative unit. The reference is to two regions in the south that constituted bases for the line of frontier fortifications, *Saltus Gerarticus* and *Saltus Menois*. To this region, which is connected to the *Limes* district, we should add Zoar, which, according to Anastasius of Gaza, included grazing lands belonging to the treasury, which were worked by captives after the Arab conquest (*PG* 89:745a).

An inscription discovered in Beersheba mentions taxpayers and *Limes* soldiers from Zoar (Alt 1921; Negev 1990), indicating that not all of the settlement was a royal estate. Zoar is mentioned as one of the settlements of the *metzikim* (*Sifrei*, Deuteronomy 357 and parallels). It should be recalled that in this period as well, owners of private land were present alongside the *metzikim*. One of the former, Babatha, left a number of legal documents that were discovered in the Judean Desert caves (second century C.E.). These documents include property declarations and writs stating the names of a number of private landowners. One of the writs also mentions a royal estate that was given over to a tenant. Its location is unclear, but it, too, was probably situated in this part of the Dead Sea area (Cotton 1994; Cotton et al. 1997: No. 60). 'En-Gedi is an additional example of the combination of royal estate and regular settlement. The settlement is called "the village of the master," i.e., the emperor (Lewis 1989: No. 11 and others), but independent property holders resided in the same settlement. The same document that contains the term "village of the master" describes a loan taken by one of the settlement inhabitants from the commander of the military unit encamped there. Other documents mention landowners, a marketplace, and an army camp. This town was therefore an estate inhabited by tenants, as well as by independent landowners, some of whom were quite wealthy.

A "king" who grants land to tenants appears in a number of midrashim (such as Genesis Rabbah 49:12, p. 499), and such grants apparently were a well-known and common occurrence. The villages of the king that are mentioned were not regular estates. Both 'En-Gedi and Prophirion were governmental centers rather than villages (for 'En-Gedi, see above). Prophirion contained a *boule* (council) and episcopate, which constitutes proof that the site included a population of independent citizens, and possibly also private landowners, who lived and were active alongside the royal estate (Honigmann 1939-1944). This complex legal structure will be discussed further below.

Lands of the *Polis*

The Roman *polis* possessed lands that were divided into three types: public buildings, profitable public installations, and lands that were a source of revenue. The first category, which is the least relevant to the current discussion, included the plots on which public structures had been erected. These lands did not generate revenues; on the contrary, their maintenance was very costly for the city. The second group comprised a small number of public installations that were capable of yielding profits. The most important of these was the bathhouse, which could either have functioned as a free municipal service, or leased to an inhabitant who collected an entrance fee for his livelihood. Additionally, the city could have had profit-generating workshops (see below). These two categories are of no agrarian importance, because the lands on which the installations had been erected no longer were of any independent economic worth. The third type, and the most important for our purposes, consisted of lands in the built-up area and agricultural plots. These were leased to entrepreneurs, and the revenue they produced constituted one of the city's primary sources of income (Jones 1966: 25-241).

In the early Hellenistic culture, city lands were a common phenomenon. Athens and Delphi possessed extensive lands, and many extant documents mention the regulations governing their management. The lands were leased to the citizens of the city for a minimal rental fee. The agreements were in force for a limited time, generally a decade, or for longer periods of time (30 or even 40 years). Similar agreements were drawn up for the leasing of a theater or a quarry (Osborne 1988). The leasing of a theater could have been a source of livelihood, or a contribution to the city. The rents were extremely low, and Osborne notes the clear lack of effort to make maximum use of the land. The sources offer no explanation for this, nor is it stated explicitly. The low sums may have been a consequence of the situation in which transactions were conducted between the city officials, who belonged to the local elite, and other members of this same elite. An additional possibility is that this constitutes further testimony to the economically inefficient manner in which public property and transactions were managed. This state of affairs has already been noted above, and will be discussed further below.

A number of early Hellenistic inscriptions discovered in Babylon tell of the land grants made by Seleucus I to the city. The inscriptions imply that the city already possessed lands, which were leased to the citizens of the city (Sarikisan 1969). Hence, the concept of the management of city lands had already been brought to the East by this period, and apparently was known in Babylonia even before the Hellenistic conquest.

Direct testimonies concerning city lands in Palestine are few in number. A single passage in the Talmud makes mention of shops that generate rent for the "*medinah*," that is, for the *polis* (T Avodah Zarah 7:1), but it transpires that this phenomenon was common in Palestine, as in other provinces.

Lands of the Rural Community

The rural community throughout the Empire was less developed than that in the *polis*. Nevertheless, the former operated public workshop installations, to which there are testimonies from Syria and from Egypt. Inscriptions from Syria, for example, commemorate the establishment of a workshop in 386 CE (Prentice 1908: No. 734; Waddington 1870: No. 2309), the construction of a public dovecote (Prentice 1908: no. 758), or a public oil-press (Prentice 1908: no. 152). The testimonies are numerous, and this phenomenon is well-known (Harper 1928). In Egypt, where the rural community was less well-developed, we also know of the public dovecote towers in Kerkeosiris, in Karanis (Husselman 1953; Tebtunis Papyri 1:79:81; see also Crawford 1971), and of additional workshop installations (e.g. Theodorides 1982).

Within this context, the community also possessed the lands on which the installations were built, as well as additional lands that were leased out. The two inscriptions from east Syria quoted by MacAdam (1984) speak of land that belongs to the village and was leased to members of nearby tribes. Regardless of the method of working the land employed by the tribesmen, the land clearly belonged to the community, which leased it to entrepreneurs. Additional testimonies to this were published by Harper (Harper 1928:151-52). In Egypt, with its less sophisticated communities, some villages owned public lands that were leased out (Gagos and Van Minnen 1992).

Lands not under cultivation were generally the property of the settlement as a whole, and were subject to the supervision of the settlement council. The rural law book published by Ashburner states this expressly (e.g. Ashburner 1909: clause 82). Anyone seeking to build on the land or to plant it had to come to an agreement with the village, and the council was naturally responsible for the payment of taxes on these plots, which had served for pasture until then (Brand 1969). We may conclude that in the adjoining provinces the rural community possessed lands, some of which were intended for the construction of public installations, and others leased to private individuals, or perhaps worked by the public authorities.

The structure of the rural community in the Jewish sector of Palestina was different. On the one hand, the community was extremely well-developed, to a much greater degree than those in the other provinces. On the other hand, it did not collect taxes, nor did it construct or operate public installations. Despite the wealth of sources, there are no testimonies relating to public workshop facilities. We hear of service installations such as a public bathhouse, a ritual bath, or similar facilities, but not of workshops. The only possible testimony is a decorated mosaic pavement uncovered in Huldah, which includes Jewish symbols and a dedicatory inscription commemorating the donations by individuals within the community. This inscription is commonly understood as evidence of a synagogue. Kloner, however, showed that this was most probably a wine-press (Kloner 1986), albeit without successfully explaining why a wine-press merited such a magnificent pavement. If Kloner is correct, then this is the sole evidence of a productive installation belonging to the public. The paucity of evidence is likely to seem insufficient proof that there were no such installations. The lack of evidence is striking in comparison with the plethora of evidence regarding the activities of the Jewish community (Z. Safrai 1995). Furthermore, there is supporting evidence that this absence of findings is not incidental.

Another *halakhah* determines that charity collectors who are left with a surplus from some appeal may not invest this sum in any commercial enterprise, even if abstaining from doing so is liable to cause a loss (Z. Safrai 1995:307). The Rabbis also opposed the public's becoming an economic factor (see below). Such an approach can easily be understood at present. It is axiomatic in a modern economy that the public may not act as a capitalistic element in the marketplace. Such activity invites corruption, and is by its very nature inefficient. In a small economy, a representative of the public would clearly have difficulty in contending and competing with the economic interests of the wealthy inhabitants of the city. We cannot determine to what degree the Rabbis were aware of such theories, but their opposition to economic activity is quite unambiguous. It may therefore be reasonably argued that the Jewish village possessed no lands under public ownership. The status of land unsuitable for cultivation, such as rocky ground and pasture land, and the status of bodies of water, will be discussed below.

Temple Lands

The lands of sanctuaries were a central component of the agrarian structure in the East in the Biblical period, as well as in the Hellenistic-Roman period. There were a great many temples. A regular *polis* contained at least four or five sanctuaries, to which we must add the rural sanctuaries, on the one hand, and the regional sacred centers, on the other. The temples received generous contributions, and their direct expenses were not great. The majority of the public was devoted to its religion, and donations of money or land were naturally a common phenomenon. Moreover, many families, especially those of the wealthy, did not have any sons. They generally sought to raise a single male heir, but high infant mortality often upset all these plans. This voluntary reduction of births at times left the family without an heir. In such a case, the contribution of land to the sanctuary was the norm. On the other hand, the temples did not sell land, since they had no pressing economic need to do so, and such sales would have run counter to accepted practice in their world. All this led to the accumulation of lands by the ancient temples.

As a general rule, the temple did not work its lands, but leased them to small-scale tenants or to members of the elite, who would lease the land to subtenants. The economic management of the temple was capitalistic in every sense, and inscriptions commemorate the activity of Apollo or Artemis as though they were accomplished businessmen. In some instances, the estate of the sanctuary was managed as a private estate, in others, as a royal estate; these lands occupied an interim position between private and public land (Rowlandson 1996:61-2; Isager and Skydgaard 1992; Linders and Alroth 1992). Another type of temple land was the hallowed forests whose use was not permitted, and which therefore had no direct economic worth.

With the adoption of Christianity by the empire, began a gradual process of withholding resources from the pagan temples. Theoretically, the Church acquired the property of the temples, in a process that requires an extensive analysis that would exceed the scope of the present discussion. At any rate, Church property is already mentioned in the early fourth century, and the legislation of Constantine implies that the Church possessed real estate even before it became the official religion. In relating to this issue, Eusebius speaks not only of churches but also of gardens (Eusebius, *HE* 10:15-17). A decree issued in 320 C.E., less than a decade after the empire became Christian, already mentions shops belonging to the Church, and its commercial dealings (*Ibid.* 9:7-11). The property of the Church would later increase as a result of contributions by individuals, the infusion of government monies, and the expropriation of the properties belonging to the temples. The monasteries possessed lands that were worked by the monks themselves (see below). In addition, the Church gave over its lands to lessees or tenants, and a special official acting on behalf of the Church or the monastery (usually known as an O*ikonomos*) was responsible for such matters. Church lands were exempt from taxation. Thus, for example, the Church of St. John in Ephesus had such an exemption, which also applied to the lands of the lessees and the *colons* (FIRA 1:466-67), independent farmers who received lands as tenants and were bound to the land (see below). The language of the law indicates that Church properties were also given over on a tenancy basis, and not only leased. The Church did not customarily sell lands, either, and thus it accumulated a great deal of real estate (Codex Justinianus 1:2:17).

It is extremely difficult to determine the extent of Temple-owned lands in Judea during the Second Temple period. According to Biblical law, land dedicated to the Temple may be redeemed by the person who dedicated it or by another, and in the Jubilee year the field returns to its original owner (Leviticus 26:14-28). In only one instance does the land remain in the possession of the priests: "But if he does not redeem the field, and the field is sold to another when it is released in the Jubilee, the field shall be holy to the Lord, as a field proscribed; it becomes the priest's holding" (Leviticus 27:20-21). Thus, there is a possibility that the plot will remain in the possession of the priest; this passage also alludes to an additional law, that pertains to land that has been "proscribed" on behalf of the Temple Regarding such a situation, the Scripture states: "But of everything that a man has proscribed for the Lord [...] nothing may be sold or redeemed; every proscribed thing is totally consecrated to the Lord" (*ibid.*:verses 28-29).

Questions regarding the sources of the law, when it originated, and whether it was implemented in the First Temple period, are secondary. The most important datum for our purposes is that this Biblical passage was known to those who established the law in the early Second Temple period and later. The interpretation of these verses by the Rabbis appears in M Arakhin and its parallels (M Arakhin 7:3-5). The Tannaitic *halakhah* related to the laws of consecrated property, both when the law of the Jubilee year was in effect, and when it was not applicable. The law of the Jubilee year was not observed in the Second Temple period, and therefore the discussions concerning the status of consecrated property when the Jubilee year is observed are of a theoretical nature, and do not attest to an actual reality.

In the language of the Mishnah, when the Jubilee year is not observed: "If one consecrated his field when the law of Jubilee is not [applicable]" (M Arakhin 8:1; T Arakhin 4:22), the one who consecrates is permitted to redeem the field that he had so designated. If not, then the field is put on sale. The Mishnaic text does not state what happens if there are no purchasers for the field, or if the Temple treasurer thinks that the offered price is too low. Another option is to dedicate a field on behalf of the Temple. According to the Mishnah: "Things dedicated for [use by] the priests may not be redeemed, but are given to the priests (as *terumah* [heave-offering]). Rabbi Judah ben Bathyra says: What is dedicated without specification is for Temple maintenance" (M Arakhin 8:6). The disagreement is whether a field that has been dedicated is given to the priests or to the Temple fund; in either event, this is plainly a possible source for providing lands for the Temple.

Another potential source of Temple lands is securities. If a person owed money to the Temple and did not pay, the Temple treasurer was authorized to collect this sum from the defaulter's property. The principle that "that belonging to the Temple has the advantage" (M Shekalim 4:9; 5:4; Arakhin 9:2, and more) was applied in many spheres. Yet another possibility, the dedication of trees, or of the fixed dedication of fruits, so that theoretically the land would remain nonsacred property, will be discussed below. All of these possibilities are in accordance with the Pharisaic *halakhah*.

There is no certainty, however, that this was the sole reigning *halakhah*. The views held by other sects have not been determined, nor has the extent to which they could influence Temple practice. Furthermore, all the *halakhot* quoted from Rabbinic sources are later formulations, and we cannot be sure that they had already been accepted when the Temple still stood. It is plausible that the Temple leadership wished to purchase lands, and it has not been determined to what degree it was limited by these *halakhot*, which are familiar exclusively from later literature. Moreover, the *halakhah* does not specify a time limit for the redemption of the consecrated field, so the Temple treasurer could delay the redemption indefinitely, until the advent of the Messiah.

We may therefore conclude that the Biblical law, and later the Talmudic *halakhah*, attempt to limit the possibilities of land purchases by the Temple. In practice, however, opportunities for receiving private lands on behalf of the Temple remained, and the Temple was not forbidden to purchase lands and to maintain possession. In antiquity, the temples were among the major landowners, and it could therefore be expected that the Temple in Jerusalem, which was especially wealthy, would also own lands.

Second Temple period sources contain few testimonies relating to Temple lands. Some of the information pertains to a field held by the Temple only temporarily, until its redemption, but mention is also made of the harvest of a field that is Temple property and of lands consecrated to the Temple, which for some reason were not immediately redeemed. The Mishnah discusses one who consecrates: "... a tree, and it later bears fruit, a field, and it later produces herbs" (M Me'ilah 3:6). This *mishnah* continues to discuss the harvest from Temple property, in connection with the laws of a spring issuing from a consecrated field, and palm trees belonging to the Temple, from which the elders take *lulavim* [one of the Four Species] for *Sukkot* (*ibid.*:mishnayot 6-8; T Me'ilah 1:19-25).

Consecrated fruits and field crops, as well as consecrated dovecotes or caves, are mentioned in many sources. The commentaries, as well as the Talmuds, generally speak of fruits that were consecrated after being picked, or shortly before the field was redeemed. These interpretations - i.e. that the land itself was not consecrated - are based on the rarity with which consecrated lands are mentioned. Such interpretations, however, are not conclusive, and these *mishnayot* may be interpreted as referring to fruits that grew in a consecrated field. A number of *halakho*t refer to permanently consecrated land: "If the roots of a privately-owned tree extended into consecrated [ground]" (Me'ilah 3:7); the same law applies to the grove of palm trees from which the elders took *lulavim* (*ibid.*). This is not a description of an event, but of a standard practice. In addition, in the *halakhah*: "Laborers who work in the sanctuary may not eat dried figs that are consecrated [...] And so: a cow that was working in the sanctuary" (T Me'ilah 1:21). Judging by the context, the laborers are engaged in fruit picking, and the cow in threshing, or even in plowing - in a consecrated field that belongs to the Temple.

The Jericho region seems to be indicated as one of the areas containing consecrated lands. The Mishnah relates that the people of Jericho would eat "*gamaziyot* belonging to the Temple"; the Rabbis protested against this, but to no avail (M Pesahim 4:9; T Pesahim 2[3]:19; BT Pesahim 26a-b; Buechler 1912:121-35). *Gamaziyot* are the fruit of the sycamore; the BT also mentions in this context carob trees from which the men of Jericho ate. The reasoning of the inhabitants of Jericho was: "Our fathers said to them, they consecrated only the trunks, because violent men would come and take them by force" (T Pesahim 2 [3]:19; T Menachot 13:20). The question of why they set aside sycamore trees will not be discussed in this article. Of importance to the current discussion is the fact that these trees had been Temple property from ancient times. Also of interest is the method of consecrating trees without the land, so that theoretically the land would not be Temple property. This method was possibly chosen in order to overcome (Pharisaic?) *halakhic* opposition to Temple ownership of lands. The description implies that this passage refers to quite extensive areas of sycamore trees, and that the custom was practiced by many of the region's inhabitants: "the people of Jericho," and was not an isolated incident.

The connection between Jericho and the Temple may also be inferred from the (exaggerated?) tradition that twelve watches (clans), about half of all the priests, lived and were active in this city (BT Taanit 27a; Buechler 1966:120-21). The *baraita*, as it is transmitted in the BT, explains in detail that when the turn of each watch came to serve in the Temple, part of the watch would go up to Jerusalem, while another part would go to Jericho and provide "water and food for their brethren in Jerusalem." Buechler had doubts regarding this passage, and believed that it was solely a consequence of the internal dialectic of the discussion in the BT, since the description is given in response to the remark that if there were 24 watches throughout the Land and 12 in Jericho, then there were more than 24 watches in all. This issue has no historical background. The description of the journey to Jericho does not constitute a response to this objection, but is independent, and constitutes a continuation of the *baraita*. The entire tradition is therefore to be read as follows (the additions by the Babylonian Amoraim, which are not an integral part of the *baraita*, are in parentheses):

> Our masters have taught: There were 24 watches (clans) in the Land of Israel and 12 in Jericho. (Then there were more? Rather, 12 of them were in Jericho.) When the time came for the watch to go up, half of the watch would go up from the Land of Israel [variant readings: to the Land of Israel], and half of the watch would go up to Jericho [from Jericho] to provide water and food for their brethren in Jerusalem.

The wording "water and food" is unquestionably an exaggeration, because water was not brought from Jericho to Jerusalem. The Jericho region, however, was perceived as a major supplier for the Temple. Our proposed interpretation, that estates belonging to the Temple were situated in Jericho, is supported by this passage, which means, in our view, that some of the priests would go to the Jericho region to work on the Temple estates. Also mentioned above are consecrated dates, which were probably brought from the Jericho area; it is possible, however, that they were consecrated as fruits that had already been picked.

An additional vague allusion may be added to this hypothesis. The Tannaitic tradition transmits an *aggadah* about the descendants of Benjamin, who gave an area of 500 X 500 cubits from their landholdings for the construction of the Temple, and in exchange received an identical area in Jericho: "When Joshua apportioned the Land of Israel, he set aside a fat pasture ground of Jericho, 500 cubits by 500 cubits, and gave it to the descendants of Jonadab son of Rechab the best part [...] and when the Divine Presence rested in the portion of Benjamin, they stood and vacated it" (*Sifrei*, Deuteronomy 352:12, pp. 411-12 and parallels).

This is plainly aggadic, but it may possibly be rooted in the reality that this area, "a fat pasture ground [*dushna*] of Jericho," contained an estate belonging to the public and to the Temple. The term "*dushna*" is of interest, since in the Aramaic of the Persian period such a term refers to royal land that was given over to a senior official to be worked, and the term also appears in the Rabbinic literature (Szuben and Porten 1987). This may allude to the fact that in the Jericho area there was a "*dushna*" of the Temple that was given over to tenants, namely, the priests of Jericho.

Based on the seal impression of Ario, who was the Temple treasurer, Avigad claims that an estate of the Temple was situated in Jericho. This hypothesis corresponds well with the series of explanations we have proposed, based on later sources, and the various conjectures combine to form a complete and logical picture (Avigad 1958).

A similar seal of Ario may also have been discovered in Ramat Rahel. An estate belonging to the Temple may have been located in Ramat Rahel as well, but this conjecture is doubtful on epigraphic grounds.

It may be concluded from the sources cited above relating to the early and late Second Temple period, that the hypothesis concerning Temple lands is supported by a considerable number of testimonies. Nevertheless, these areas were evidently not extensive, and the Pharisaic *halakhah*, at least in its later version, did not look favorably upon the proliferation of lands owned by the Temple. Since the possession of lands by a sanctuary was common in antiquity, and the Temple in Jerusalem was extremely wealthy, the paucity of testimonies regarding Temple lands in Judea requires an explanation.

It may reasonably be assumed that if the Temple owned lands, then upon the Hasmoneans' ascent to power, the priestly and royal authorities in Judea would have been united, and the properties of the Temple would have been identified, at least partially, with all the properties of the kingdom. This process may have begun as early as in the Persian period, when the province of YHD was granted extensive autonomy. Already in this period, the High Priest was incorporated into the autonomous civil administration, and it is therefore conceivable that the distinction between the properties of the Temple and those of the secular government had already began to blur.

As noted above, the seals attest to the inclusion of the Temple properties under the name of the province. The first half of the imprint mentions the name of the province (YHD), and the second half alludes to the Temple possessions and the treasurer in charge of them (Ario). This imprint seems to indicate the inclusion of Temple lands within the framework of the secular government. This hypothesis is supported by an analogy known to us from Asia Minor in the well-known inscription from Aezani in Phrygia (CIL 3:355). The episode it relates is as follows: An unidentified Hellenistic king expropriated the lands of the temple of Zeus in the valley in order to settle people there, while allotting to the city half of the rental sums to be obtained from these lands. It is highly likely that the jars in Judea bearing the seal impression reflect a similar arrangement. In other words, the Temple lands were expropriated and transferred to the state treasury, while the Temple retained partial rights to the land or the crops. In any case, the conclusion to be drawn from this interpretation, that the Temple in Jerusalem possessed lands, seems reasonable and logical.

The pagan temples in Palestine owned lands, as did any normal sanctuary in the Roman East. There are no direct testimonies to this, but a temple without land is hardly conceivable.

Consecrated forests were probably situated around many sanctuaries. It was forbidden to cut down trees, to graze, or to plant in these groves. Once again, there are no testimonies from Palestina, but such a situation is extremely plausible. Gabel Madbach above Petra was a sacred site, reached by two staircases, one probably for ascending, and the other for descending. Beside each of the staircases was a small temple, known today as the "garden temple," alongside which was an irrigated plot of land extending over an area of approximately two dunams, apparently a sacred grove.

In the Jewish sector this phenomenon was more limited. The attitude to sacred sites outside Jerusalem was mixed (Z. Safrai 1987). Nonetheless, there are reports of a similar forest. The inhabitants of Migdal ask a sage concerning a grove of acacia trees from which, tradition asserts, lumber was cut for the Temple. According to the narrative, the neighboring inhabitants regarded the forest as sacred, and most likely refrained from making any use of it; in other words, this was consecrated property that could not be used. The sage does not sanction the custom, but determines that the ancient practice is to be continued (PT Pesahim 4:30[d] and parallels). Accordingly, this phenomenon is not unknown, although possibly not as common throughout Judea as it was in the pagan realms.

In the Byzantine period, the Church became an agrarian-economic factor. The reports from Palestine are quite clear. Many monasteries in the land, as well as other monasteries in the Byzantine world, were economic entities. These monasteries generally contained oil presses and wine presses, which attest to the large-scale production of oil and wine. Such monasteries include those in Shelomi, Kursi, Modi'in, and very many others. In all of them the monks themselves engaged in agriculture, and the monastery functioned as an autonomous agricultural settlement. The books of responsa by the monks in the south contain various queries pertaining to church and monastery property, commerce, and the transport of goods, as a regular commercial body. Cyril of Scythopolis also tells of lands belonging to monasteries (Cyril of Scythopolis, *Vita Sabae*, chap. 51), or of a certain person named Mamas who donated money and the revenues from his lands to the monastery of Sophronius (Cyril of Scythopolis, *Vita Theododosii*, chap. 5), and there are additional similar testimonies.

Some of the monasteries in Palestine included several agricultural installations. At the same time, the Church owned lands that it leased to tenants. Marcus Diaconus, the author of the memoirs of Porphyrius (the Bishop of Gaza in the fifth century) tells of Baruch, the bishop's assistant, who collected monies from the tenants of the Church in the region. According to this description, the tenants, or some of them, were not Christians. They regarded the Church as a harsh and evil landlord, like every similar landlord (Marcus Diaconus, *Vita Porphyrii* 22; Dan1976: 183-88).

The Lands of the Legion

At times, ownership of the land was transferred to the Legion, with the revenues intended for the Legion's expenses. The Legion did not work the lands itself. Like all imperial lands, they were given over in tenancy or leases. The area was parceled into square plots, each of which was given to a different centurion, and from which payments were collected on behalf of that centurion. Throughout the Empire, stones bearing inscriptions characteristic of the boundary markers of such divisions have been discovered (Kandler 1975; Schulten 1984). A similar stone was discovered in the Jezreel Valley, and this entire region was in fact divided into square plots, as was the practice with Legion property. One stone is not sufficient, however, and the organized division of the land may also be given a different interpretation (Isaac and Roll 1982:105-6; Z. Safrai 1994:349). In one instance a centurion is mentioned as a tax collector (Sperber 1969). This presumably constitutes proof that this land belongs to the Legion, because only under such circumstances would an army officer be likely to function as a tax collector. This proof, however, should be treated with caution, because at times the sources are not precise in their use of terminology, and it may be presumed that in a rural region an army officer also performed the duties of a tax collector, or served as an escort for such a functionary. For the Jewish farmer, such a distinction between the two was vague.

According to an inscription discovered in Syria, a boundary marker of the village lands was erected by order of an army officer. Since the person erecting the marker was an officer, Prentice surmised that this referred to land belonging to the Legion (Prentice 1908: No. 666). Once again, however, the distinctions between the military and civil authorities could have been obscured in a border region, in which there was a strong military presence. Although this is an official inscription, rather than a literary portrayal, it is doubtful whether this is sufficient proof. The surveyors of Hauran and eastern Syria did not find proof of an orderly Roman land parceling, but rather of less orderly, earlier division of land. The element of Legion land was therefore marginal in this region.

It is also known that on occasion, Legion soldiers would work lands adjoining their camps, or graze their cattle there. This was not formal ownership, but a local arrangement (MacMullen 1963:1-22). There is no such evidence in the Land of Israel.

The Lands of the *Limes*

A new type of land ownership by soldiers developed in the fourth century. This century, approximately, marked the establishment of the *Limes*, about which much remains to be learned. The southern region was problematic, in terms both of settlement and of security. A certain military presence, the details of which are subject to scholarly debate, may possibly have existed here before this century. Scholarly opinion, however, is unanimous that in the early fourth century, during the reign of the emperor Diocletian, a policy change was effected, when the Tenth Legion was transferred from Palestina to the Negeb, and the Sixth Legion from Galilee to the eastern Jordan Valley (there is much literature on this, see: Gichon1975; 1991; 1999; Schatzman 1983; Z. Safrai 1994b and additional literature therein).

The scholarly literature states simply that the military presence was accompanied by special agrarian arrangements. The soldiers received plots of land, which constituted one component of their wages. According to this description, the military disposition was a settlement enterprise par excellence, on an extremely large scale. Consequently, a detailed examination is necessary to do this subject justice.

The *Limes*, as a military settlement undertaking, is known from the middle- Byzantine period. According to contemporary testimonies, lands were allocated to the soldiers, with each one probably receiving a small estate, in exchange for his army service. Upon his retirement, his son entered the army in his stead, and from then on, the estate always had to provide the army with a single recruit. The entire region became an administrative unit directed by the commander of the Legion, who assumed the role of both a military and a civil commander. The duty of serving was initially incumbent upon the son and heir, but this son would later be permitted to send a replacement from among the estate's tenants or workers. At an even later stage, the institution of the *Limes* degenerated, and instead of providing a soldier or an officer, the estate was required to pay a monetary tax in the amount of a soldier's wages. This arrangement developed gradually, and eventually became solely financial (Haldon 1979; Price 1974).

Although scholars believe that this *Limes* arrangement already existed in the early Byzantine period, such an assumption is not simple. First, there is no certainty that the area of the military deployment was formally called the *Limes*. This term appears in a number of testimonies beginning from the fifth century, but these are literary expressions, and it cannot be proven from them that this was an official legal name for the region already in this period. Moreover, there is no certainty that the arrangement known from the late seventh century was already in existence three centuries previously.

On the other hand, a number of testimonies support the argument that some form of the *Limes* arrangement was in effect in the Negeb at least in the sixth and seventh centuries. Many papyri of soldiers in the Byzantine army were found in Nessana, indicating that these soldiers were the owners of lands that they also sold and purchased (Kraemer 1958). Moreover, an analysis of the papyri and the administrative inscriptions discovered in Beersheva indicates that in the late Byzantine period the soldiers stationed in Nessana, as in a number of other settlements in the region, received generous monetary grants. Later on, under Muslim rule, they became taxpayers to the new authorities (Negev 1990).

In contrast, there is no evidence of agricultural activity in the small outposts and camps. Most were not suitable for agriculture, and soldiers stationed in them were probably not given lands. An oil press was discovered at only one outpost in Kh. el-Qasr (Kloner and Hirschfeld 1987). Even in this case, however, it is not clear whether the press was a part of the camp, or whether the owners took advantage of the security provided by the small camp and built the oil press next to it. Moreover, it is possible that the site is not a *Limes* camp, but a regular camp or even a fortified farm dating from the first century.

In the other small forts there is no hint of extramilitary activity, nor was there any possibility of working the desert terrain, as in Ein Boqeq, for example,. An agricultural system was discovered near Ein Boqeq, including an aqueduct and an installation, but the system is not contemporaneous with the fort (Gichon 1993). In addition, next to Mezad Tamar (Thamara) there are no agricultural lands, as is true of other forts. The situation is Jotapata is less clear. The large camps in Transjordan contain evidence of limited civilian settlement, but, again, it has not been determined whether this dates from the time of the camp, or was a secondary settlement, established after the camp had been abandoned (Parker 1986; 1992. For a general discussion of the problem, see Gichon 1999).

All these testimonies indicate that the *Limes* had a complex agrarian-economic structure. The soldiers apparently were the recipients of generous plots of land that enabled them to establish small estates and to become wealthy. The plots were centered around the large settlements, where the headquarters of the military force were situated. There were no such plots of land next to the forward camps. The soldiers were stationed in these camps on a rotation basis: they spent a part of the time in the vicinity of the headquarters, where they could closely supervise their tracts of land, and the rest of the time performed their duties in the garrison force in the forward camps. During times of crisis they were dispatched to the forward positions; it is also possible that other soldiers, who were not granted lands, served in these positions.

In administrative terms, the *Limes* was considered to be an independent district. The administrative lists from the sixth century mention *Saltus Gerarticus* and *Saltus Menois* (Avi-Yonah 1963:168-70), and the original meaning of the term *Saltus* is government estate. In the sixth century this term referred to the area of the *Limes*, which in practice became the plots granted to soldiers, which later became the private property of these soldiers.

After having clarified the legal and military background, we must now examine the settlement-agrarian significance of this arrangement. The settlement growth in the southern Hebron hill country in the fourth century is most likely linked to the establishment of the lines of fortification in the northern Negeb and the eastern Hebron hill country. It may be surmised that security problems made it difficult to develop a settlement infrastructure in the frontier regions. The presence of the army facilitated matters for the inhabitants and improved the security situation in the vicinity of the outposts, on the frontier, and in the Hebron hill country. This was of decisive importance in the first stages of the settlement's development, before it had gained strength as an independent entity. The army may have been assigned to the area for various reasons, but the move resulted, perhaps indirectly, in increased security in the frontier region. There may have been other reasons for transferring the army to the territory. Above, we raised the hypothesis that the establishment of the administrative framework of the *Limes* was a result of economic and administrative considerations: the emperor wanted to grant lands to his soldiers and to create a stratum with a large and fixed recruitment potential. Vacant lands came into the possession of the emperor mainly on the fringes of the settled part of the land, and this may have been why military units were transferred to this region.

It is difficult to determine the extent to which security factors influenced the history of the frontier settlement, but the timing and chronological synchronization of the two phenomena cannot be ignored. The enhanced military presence and the construction of a line of military fortifications, on the one hand, and settlement growth, on the other, occurred simultaneously in the early fourth century. Additional aspects of the link between these two phenomena will be discussed below.

The *Limes* region and the settlement in the northern Negeb and the southern Hebron hill country warrant a detailed examination. Sufficient for the current discussion is the brief conclusion that the transferral of the army to the southern Hebron hill country influenced the settlement picture in this region. This settlement activity was comprised of three elements:
(1) the lands in the area were given the special status of *Limes* lands, and the settlers were granted a sort of subsidy (military salaries).
(2) the entire region enjoyed improved security.
(3) the transferral of the army created a focal point of consumer demand that aided the economy of the entire area.

Ownerless Lands - Desert and Rocky Areas

Every settlement in the hilly regions of Judea-Palestine contained areas that could not be cultivated. These areas were as limited as possible, because the inhabitants in antiquity made great efforts to utilize the land to its maximum. Despite the labor invested to this end, there still remained rocky patches and steep slopes that could not be cultivated. The question that arises is whether these areas were allocated to private holdings, were regarded as public lands belonging to the inhabitants of the city, or were ownerless lands. The next question that naturally must be addressed is whether an entrepreneur could have taken control of such an area and turned it into a private holding.

In Roman and Byzantine law, abandoned land was defined as land for which taxes were not paid (Brand 1969). This definition included fallow private land. Any farmer was entitled to take over abandoned land, which would become his if he paid taxes on it, including harvest taxes. In other words, the working of abandoned land led to ownership of it. The law was meant to encourage the working of the land, i.e., the payment of taxes, and to penalize anyone who did not honor his obligations to the state treasury. Those primarily harmed by this law were the poor, who were incapable of meeting the tax payments.

Such a law does not exist in the Jewish legal system. According to the *halakhah*, "land is not stolen," i.e. the rights of the owners cannot be expropriated in any manner without their consent. The question naturally arises as to the status of lands that had never been worked, or had been abandoned by their owners.

The rural laws published by Ashburner clearly state that uncultivated land belonged to all the villagers; in other words, this was land that had not been allocated. Anyone seeking to work it had to receive the approval of the council, and may also have been required to pay this body. If he did not do so, then the cultivated land, or any structure built upon it, belonged jointly, in equal shares, to the council and the entrepreneur (Ashburner 1909:Section 81). This policy was a distinct deviation from the rules, according to which any entrepreneur could develop fallow lands and even enjoy tax concessions in the first years (see below). In Egypt, for example, uncultivated (in practice, non-irrigated) lands returned to the possession of the authorities, who could lease them out as they saw fit (Rowlandson 1996:50). The late Byzantine law ruled that such lands would be given over to the lessees, who would gain ownership of them after 30 years (Brand 1969). Needless to say, implicit in the language of the law is a subsidy to the lessees, which was intended as an incentive for the (renewed) use of neglected lands.

The Rabbinic literature contains two or three terms relating to untillable land: *terashim* (rocky ground), *horashin* (thickets), and *midbar* (wilderness). The references to these regions are incidental and minor. The Tosefta teaches: "They remove stones from a road in the public domain; this is the opinion of Rabbi Joshua. Rabbi Akiva says, Just as no one has the right to disrupt the public way, so no one has the right to remove stones. But if one has removed stones, he should take them out to the sea or river or stony ground [*terashin*]" (T Bava Kamma 2:12). The problem was finding a place to which the stones could be removed. The solution of transferring them to stony ground is straightforward in the legal sense, but extremely costly. On (already) stony ground the stones cause no damage. One could understand from this that *terashin* is an ownerless area, and therefore superfluous stones could be cast there. Also possible, however, is that formally, these stony areas were privately owned. The owner, however, could not object to the placing of the stones on his property, because they cause no harm and, in the final analysis, some place had to be found for them. The public interest necessitates finding a solution, and this need may be stronger than the right of private ownership. Similar legal solutions will be discussed below.

Another *halakhah* is concerned with one who sells an estate house ("*ir*" in the language of the sources). According to the *halakhah*, the sale includes all the agricultural areas and the structures that are an integral part of the estate, but not external structures: "Even though he [the seller] said to him [the purchaser]: 'It and all that is in it, I sell to you' - he did not sell him the outlying parts or suburbs, or the thickets [*horashin*] that are set apart by themselves" (T Bava Batra 3:5; cf. BT Bava Batra 69b; PT Bava Batra 3:14[c]). Therefore, *horashin* are not part of the agricultural plot, but are connected to the estate, since they are "set apart" from it [i.e., designated for someone specific, while part of the estate]. The probable intent is that the estate owner is entitled to make use of them, but does not have legal ownership of such areas. This right is not part of the estate, and therefore these areas are not sold with it. In this same context the Palestinian Talmud adds that "a share in the sea and in the river" namely, fishing rights and the use of the ownerless area, also belong to the "*ir*", and the owner of the estate house successfully claimed exclusive rights of use of the ownerless property within or adjoining the bounds of his estate. This type of claim may also apply to the *horashin*, which were ownerless, with the estate owner exercising rights beyond those of a regular farmer.

It thus seems that the *horashin* areas were generally ownerless, and farmers could, at most, claim special and exclusive rights of use. Under no circumstances were they considered part of the landholding, and their use may have been a special "right" of the powerful, ensuing from the great power wielded by wealthy landowners.

The *horesh* was generally perceived as ownerless, and therefore a vine growing in it does not fall under the laws of *orlah* (in which the fruit of the first three years is forbidden), the fruit is considered ownerless (PT Orlah 1:60[d]), and shepherds could graze their flocks in *horashin*, even in a region not their own. In the language of the Talmud: "They may graze in *horashin*, even the tribe of Judah in [the territory of] the tribe of Naphtali" (PT Bava Batra 5:15[d]). Although archaic terms are used, the intent is clear: the *horesh* has no owners, nor may anyone lay claim to it.

The Wilderness

"The wilderness" [*ha-midbar*] is synonymous with ownerless property. Thus, for example, the Torah was given in the wilderness - an ownerless area, meaning that no one owns it, and anyone may work it. Special emphasis is placed on the fact that the wilderness is exempt from the *arnon* (the *arnona* tax - Numbers Rabbah 19:15). The "wilderness" in this context plainly refers to the great wilderness of Sinai, and possibly also to the Negeb region. Also included in this term, however, is a desolate area within a settled region. Thus, for example, Rabbi Hanina ben Dosa went to the *midbar* of his city to quarry stone for a donation to the Temple (Ecclesiastes Rabbah 1:1). His going to the *midbar* is understandable, because it was ownerless, and anyone who desired to quarry there was permitted to do so.

The Rabbis were generally opposed to grazing within the settled area, but permitted such activity in *horashin* and wilderness areas (T Bava Kamma 8:10; and more). This permission was based on the fact that the ecological damage caused by grazing in the wilderness was negligible; such permission, however, would be pointless if this were not land open to all for grazing.

Abandoned Areas

Abandoned areas were common in the Roman empire. The period of revolts was conducive to the development of such a phenomenon, since at times the owners of these lands had been deported or had fled. Their lands were not confiscated, nor were they cultivated, and it was only natural that their neighbors or relatives had designs on these plots. It may be assumed that the phenomenon of abandoned lands is more prevalent in a society based on the nuclear family. A family disaster could lead to the desertion of the land for a long period of time; in an extended family, someone could take the place of the owner, which would not be the case in a nuclear family.

Another well-known phenomenon was fleeing from the land. Small landowners who were not successful in meeting the tax burden, fled from their land (e.g. Whittaker 1976). The authorities, who regarded this as a loss of revenue, permitted anyone who so desired access to this land, in order to work it and pay taxes on it. The *halakhic* situation was more complicated: the Tosefta recognizes three terms: the property of captives [*shevuyin*], abandoned [*natush*] property, and forsaken [*ratush*] property (T Ketubot 8:3). The Tosefta, and later the Talmuds, sought to explain these terms, and come to the following conclusion:
Property of *shevuyim* - an heir takes over the property of one who was taken captive (or who fled), and was reported to have died.
Property of *netushin* - an heir enters the property of one who was taken captive or who fled, and "it was not reported that [the latter] had died."
Property of *ratushin* - an heir enters the property of one who was taken captive or who fled, who disappeared and whose fate is unknown.

According to Rabbi Simeon ben Gamliel, the term "property of *shevuyin*" is the same as "property of *netushin*" (T Ketubot 8:2; PT Ketubot 8:29[a]; Yevamot 15:15[a]; BT Bava Metzia 38b; see the discussion in S. Lieberman, *Tosefta ki-Fshutah*, Bava Metzia, 314-16).

The need for the *Amoraim* to explain the terms indicates that the problem was already unfamiliar by the time of the redaction of the Tosefta, and certainly by the time of the *Amoraim*. Indeed, the Amoraitic literature contains no additional facts that would explain the terms. It would seem that the problem arose during the time of the revolts, and once again became rare in the following period of tranquility, beginning in the second half of the second century.

In practice, the property remained in the family. The *halakhic* literature attests that the court does not permit the heirs to enter the property of a father whose death has not been conclusively proved. The BT cites a disagreement on this issue (BT Bava Metzia *loc. cit.*), but sources from the Land of Israel casually speak about such a restriction (*Avot de-Rabbi Nathan*, version A, chap. 38, 58a; *Mekhilta*, *Mishpatim* 18, p. 314; *Mekhilta de-Rabbi Shimon bar Yohai*, *Mishpatim*, p. 211).

In addition to these *halakhic* definitions, a similar problem recurs in the Rabbinic literature: "If one enters his fellow's ruins and rebuilds it without permission, when [the former] goes out: if he [the one who entered] says to him [the owner], "Give me my timber and stones," his request is not granted. If one enters his fellow's ruins and rebuilds it without permission, they assess it for him, and he is at a disadvantage; with permission, they assess it for him, and he enjoys the advantage. In what manner is he at a disadvantage? If the appreciation exceed the expenditure, he is awarded the expenses; and if the expenses exceed the appreciation, he is awarded the appreciation." (T Ketubot 8:8-9; Bava Kamma 10:6-7; PT Bava Kamma 9:6[d]; Gittin 5:47[a]; BT Ketubot 80a; Bava Metzia 101a).

Accordingly:
(1) The one who enters may not withdraw his investment, because of the obligation to settle the Land of Israel.
(2) The one who enters is entitled to receive a payment from the owner, whether the "appreciation" (the amount by which the property increased in value), or his expenses, in accordance with the following criteria:
- if he entered without permission, he receives the lower sum;
- if he entered with permission, he receives the higher sum.

The BT alludes that the "permission" is that of the owner of the property. If so, however, and the owner sent the "one who enters," it is obvious that he cannot remove him; on the other hand, it is equally evident that the one who enters may not withdraw his investment, because he has invested at the request of the owner. Furthermore, if a person enters his fellow's field without permission, he is presumed to be a thief, and the *halakhah* should properly express its aversion to this crime. It would therefore seem that the "permission" is that granted by the court or by the community. All these passages clearly refer to a known phenomenon. Someone takes control of abandoned property, and this very act wins the approval of the public, which is interested in the renewal and growth of settlement life. The first *halakhah* preserves the owner's right of title to the property, while the second awards the settler compensation for his expenses. Sometimes the owner did not return, the settler profited, and so did the public. If, however, the owner returned, he regained his landholding, without causing excessive loss to the one who entered.

This phenomenon was probably more widespread than is implied by the sources; it would also seem that the permission granted by the authorities to settle on vacant land was not unattractive. The *halakhah* sought to preserve the principle that land cannot be stolen, but in practice this principle was eroded. We will return to the discussion of this struggle with the authorities and the dichotomy between the Roman law and the *halakhah* .

In conclusion, uncultivated plots within the settlement area, and certainly those in ecologically inferior regions, were regarded as ownerless, and whoever so desired could derive benefit from them, mainly for grazing. This situation created, or was likely to lead to, serious problems. One of the users could presumably overutilize the pasture land, thus upsetting the local ecological balance. There is no allusion to these problems in the Jewish sources. Either they did not arise, or for some reason they were not discussed in the extant literature, and we cannot determine which of these two possibilities is the correct one.

Notes

1 This article is part of a broader study examining the agrarian structure in the Land of Israel in the period of the Mishnah and the Talmud. The study was funded with the aid of the Land Use Institute of the Jewish National Fund and the Dr. Irving Moskowitz Chair in Land of Israel Studies in the Martin (Szusz) Department of Land of Israel Studies of Bar-Ilan University.

2 The Jubilee year cancels the sales of land and mandates that once every 50 years the land must return to its original owners.

REFERENCES

Avigad, N.
1958 A New Class of Yehud Stamps. *Bulletin of the Israel Exploration Society* 12:3-10 (Hebrew).

Alt. A.
1921 *Die Griechischen Inschriften der Palestina Tertia Westlich der `Araba.* Berlin and Leipzig.

Applebaum, S.
1967 The Agrarian Question and the Revolt of Bar Kokhba. *Eretz Israel* 8:283-87 (Hebrew).

Applebaum, S., Isaac B., and Landau, Y.
1978 Varia Epigraphica. *Scripta Classica Israelica* 4:133-59.

Ashburner, W.
1909 *The Rhodian Sea Law.* Oxford: Clarendon Press.

Avi-Yonah, M.
1946 Newly Discovered Latin and Greek Inscriptions. *Quarterly of the Department of Antiquities in Palestine* 12:86-87.
1963 *Historical Geography of Palestine from the End of the Babylonian Exile up to the Arab Conquest.* Jerusalem: Bialik (Hebrew).

Benoit. P., Milik. J.T. and, De Vaux. R.,
1961 *Les Grottes de Muraba'at.* Discoveries in the Judaean Desert II. Oxford: Clarendon Press.

Bickerman, E.J.
1979 *The God of the Maccabees: Studies on the Meaning and Origin of the Maccabean Revolt.* Leiden: Brill.

Brand, C.M.
1969 Two Byzantine Treatises on Taxation. *Traditio* 25:35-60.

Buechler, A.
1912 *The Economic Conditions of Judaea after the Destruction of the Second Temple.* London: Jews College.
1966 *The Priests and Their Cult in the Last Decade of the Temple in Jerusalem.* Jerusalem: Rav Kook (Hebrew translation of *Die Priester und der Cultus im letzten Jahrzehnt des jerusalemischen Tempels.* 1895. Vienna: Israel-Theol. Lehranstalt).

Bull, R.J.
1966 A Roman Veteran's Epitaph from Azzun, Jordan. *Palestine Exploration Quarterly* 98:163-65.

Carlsen, J.
1994 *Landuse in the Roman Empire.* Rome: L'Erma di Bretschneider.

Cotton H.M.
1994 Rent or Tax Recipt from Maoza. *Zeitschrift fur Papyrologie und Epigraphik* 100:547-575.

Cotton, H.M. and Yardeni, A.
1997 *Aramaic, Hebrew and Greek documentary texts from Nahal Hever and other sites.* Discoveries in the Judaean Desert 27. Oxford: Clarendon Press.

Crawford. D.J.
1971 *Kerkeosiris.* Cambridge: Cambridge University Press.

Cyrillus of Scythopolis
1939 *Vitae ss. Euthynii, Sabae, etc.* ed. E. Schwartz. Leipzig: Hinrichs.

Dan, Y.
1976 *Social Life in Eretz Israel in the Byzantine Period in the Sixth and Seventh Centuries.* Ph. D. diss., Hebrew University, 1976 (Hebrew).

Dawes, E., Baynes, N. H.
1948 Theodore of Sykeon. In *Three Byzantine Saints.* Oxford: St. Vladimir's Seminary Press.

Eusebius
see Klosterman 1984.

Frankel, Y.
1996 The Mamluke Sultan, Baybars, and Syria's Land Regime. *Karka* 42:84-97 (Hebrew).

Gagos, T., and Van Minnen, P.
1992 Documenting the Rural Economy of Byzantine Egypt: Three Papyri from Alabtrine. *Journal of Roman Archaeology* 5:186-202.

Gichon, M.
1975 The Site of the *Limes* in the Negev. *Eretz Israel* 12:149-66 (Hebrew).
Gichon, M.
1991 When and Why Did the Romans Commence the Defence of Southern Palestine? Pp. 318-25 in *Roman Frontier Studies 1989,* eds. V.A. Maxfield, and M.J. Dobson. Exeter: University of Exeter Press.
1993 *En Boqeq.* Mainz am Rhein: Von Zabern.
1999 The Hinterland of the Limes Palestinae – Remains and Written Evidence. Pp. 667-685 in *Roman Frontier Studies, Proceedings of the XVII International Congress of Roman Frontier,* ed. N. Gudea. Stydies: Zalau.
Grossman, D.
1996 Communal Holdings - A Discussion of the *Musha'e* System and Its Consequences. *Karka* 41:56-76 (Hebrew).
Haldon, J.F.
1979 *Recruitment and Conscription in the Byzantine Army 550-950.* Vienna: Verlag der Österreichischen Akademie der Wissenschaften.
Harper, G.M.
1928 Village Administration in the Roman Province of Syria. *Yale Classical Studies* 1:105-68.
Honigmann, J.J.
1939-44 L'Eveche Phenicien de Porphyreon (Haifa). *Annuaire de l'Institue de Philologie et d'Histoire Orientales et Slaves* 7:381-94.
Husselman, E.M.
1953 The Dovecotes of Karanis. *TAPA* 84:81-91.
Isaac, B.
1984 Judea after AD 70. *Journal of Jewish Studies* 35:44-50.
Isaac, B.H, and Roll, I.
1982 *Roman Roads in Judea 1- The Legio-Scythopolis Road.* BAR International Series 141. Oxford: British Archaeological Reports.
Isager, S., and Skydsgaard, J. E.
1992 *Ancient Greek Agriculture.* London and New York: Routledge.
Jones, A.H.M.
1966 *The Greek City from Alexander to Justinian.* Oxford: Clarendon Press.
Josephus
Jewish Antiquities. Trans. by H. St. J. Thackeray. 1989. London: Harvard University Press.
The Jewish War. Trans. by H. St. J. Thackeray. 1989. London: Harvard University Press.
The Life of Josephus. Trans. by H. St. J. Thackeray. 1989. London: Harvard University Press.
Kandler, M.
1975 Territorium Legionis von Carnuntum. *Roman Frontier Studies* 11. Limes Congress.
Kark, R.
1989 Land and Concepts of Land Redemption in Traditional Societies and Eretz Israel. *Karka* 31: 22-35 (Hebrew).
Kehoe, D.P.
1992 *Management and Investment of Estates in Roman Egypt during the Early Empire.* Bonn: Habelt.
Kloner, A.
1986 The Structure at Hulda and its use as a Vine Press. Pp. 197-208 in *Man and Land in Eretz-Israel in Antiquity,* eds. A. Kasher et al. Jerusalem: Yad Ben Zvi (Hebrew).
Kloner, A and Hirschfeld, Y.
1987 Khirbet el-Qasr - A Byzantine Fort with an Olive Press in the Judean Desert. *Eretz Israel* 19:132-41 (Hebrew).
Klostermann, E.
1904 *Eusebius - Das Onomastikon der Biblischen Ortsnamen.* Leipzig: Hienrichs.
Kraemer. C.J.
1958 *Excavations at Nessana.* Princeton: Princeton University Press.
Lewis, N.
1989 *The Documents from the Bar Kokhba Period in the Cave of Letters.* Jerusalem: Israel Exploration Society.
Linders, T. and Alroth, B.
1992 *Economics of Cult in the Ancient Greek World.* Uppsala: Uppsala University.
MacAdam, H.I.
1984 Some Aspects of Land Tenure and Social Development in the Roman Near East: Arabia, Phoenicia and Syria. Pp. 45-62 in *Land Tenure and Social Transformation in the Middle East,* ed. T. Khalidi. Beirut: American University of Beirut.
MacMullen, R.
1963 *Soldier and Civilian in the Later Roman Empire.* Cambridge: Harvard University Press.
MacMullen, R.
1990 *Changes in the Roman Empire.* Princeton: Princeton University Press.
Mann, J.P.
1983 *Legionary Recruitment and Veteran Settlement during the Principate.* London: Institute of Archaeology.
Marcus Diaconus
1930 *Vita Porphyrii episcopi Gazensis.* ed. H. Gregoire and M. A. Kugener. *Vie de Porphyre, eveque de Gaza.* Paris: Les belles lettres.
Mazar (Maisler), B.
1940 *Excavations at Beth-Shearim.* Jerusalem: Israel Exploration Society (Hebrew).
Milik, J.T., Benoit, P. and De Vaux, R.
1961 *Les Grottes de Muraba'at.* Discoveries in the Judean Desert II. Oxford: Clarendon.

Mittwoch, A.
1955 Tribute and Land-tax in Seleucid Judea. *Biblica* 36:352-61.
Negev, A.
1990 Mamphsis – The End of a Nabatean Town. *ARAM* 2:337-365.
Osborne, R.
1988 Social and Economic Implications of the Leasing of Land and Property in Classical and Hellenistic Greece. *Chiron* 18:279-323.
Parassoglou, G. M.
1978 *Imperial Estates in Roman Egypt.* Amsterdam: Hakkert.
Parker, S.T.
1986 *Roman and Saracens: A History of the Arabian Frontier.* Winona Lake: American Schools of Oriental Research.
1987 *The Roman Frontier in Central Jordan.* Oxford: British Archaeological Reports.
1992 The Limes and Settlement Pattern in Central Jordan in the Roman and Byzantine Periods. *Studies in the History and Archaeology of Jordan* IV:321-326.
Patrich, J.
1995 *The Judean Desert Monasticism in the Byzantine Period: The Institutions of Sabas and his Disciples.* Jerusalem: Yad Ben Zvi.
Petrus the Iberian
1895 *The Life of Petrus by John Rufus*. ed. R. Raabe. Leipzig: J.C. Hinrich.
Pliny (the Elder)
1969 *Natural History*. trans. H. Rackham, W.H.S. Jones. ed. LCL. London: W. Heinemann.
Prentice, W. K.
1908 *Greek and Latin Inscriptions III*. New York: Century.
Price, B.M.
1974 *The Role of Military Men in Syria and Egypt from Constantine to Theodosius II*. Unpublished doctoral dissertation. Oxford.
Procopius of Caesarea
1905-13 *Opera* (including the Anecdota). ed. J. Haury. Leipzig: Teubneri.
Rowlandson, J.
1996 *Landowners and Tenants in the Roman Egypt.* Oxford: Clarendon.
Rubin, N.
1972 For Whom Does One Mourn? A Sociological Analysis of Talmudical Sources. Pp. 111-22 in *Memorial to H. M. Shapiro*, vol. 2. Annual of Bar-Ilan University, Volume in Social Sciences, ed. H. Z. Hirschberg. Ramat Gan: Bar Ilan (Hebrew).
Safrai, S.
1965 *Pilgrimage at the Time of the Second Temple.* Jerusalem: Am Hasefer (Hebrew).
1994 The Commandment of the Sabbatical Year in the Post-Destruction Reality. Pp. 259-67 in *In Times of Temple and Mishnah.* Jerusalem: Magnesss (Hebrew).
1994 *Sikarikon.* In *In Times of Temple and Mishnah* Jerusalem: Magnesss (Hebrew).
Safrai, Z.
1983 Family Structure during the Period of the Mishna and the Talmud. *Milet- Everyman's University Studies in Jewish History and Culture* 1:129-56 (Hebrew).
1987 Sacred Tombs and Holy Sites in the Jewish Tradition. Pp. 303-313 in *Zev Vilnay's Jubilee Volume*, vol. 2, ed. E. Schiller. Jerusalem: Ariel (Hebrew).
1994 *The Economy of Roman Palestine.* London: Routledge.
1994b Between Jews and nomads. Pp. 30-50 in *Settlements and Security in the Sourthern Hebron Hills in Ancient Times,* eds. Z. Safrai and Y. Levin. Ramat-Gan: Bar Ilan University.
1995 *The Jewish Community in the Talmudic Period.* Jerusalem: Shazar (Hebrew).
1999 Ancient Field Structure – The Village in Eretz Israel during the Roman Period. *Cathedra* 89:7-40 (Hebrew).
Sarikisan, Kh.
1969 City Land in Seleucid Babylonia. Pp. 312-31 in *Ancient Mesopotamia,* ed. I. M. D'iakanov. Moscow: Nauka.
Schalit, A.C.
1969 *Koenig Herodes; der Mann und sein Werk.* Berlin.
Schatzman, I.
1983 Security Problems in Southern Judea Following the First Revolt. *Cathedra* 30:3-32 (Herew).
Schulten, A.
1894 Das Territorium Legionis. *Hermes* 29:481-516.
Spek, R.J. van der
1995 Land Ownership in Babylonian Cuneiform Documents. Pp. 173-245 in *Legal Documents of the Hellenistic World,* eds. M.J. Geller and H. Maehler. London: Warburg Institute.
Sperber, D.
1969 The Centurion as a Tax-Collector. *Latomus* 28:186-88.
Szubin, H.Z., and Porten, B.
1987 Royal Grants in Egypt: A New Interpretation of Driver 2. *Journal of Near Eastern Studies* 46:39-48.
Theodorides, A.
1982 Les Communautes rurales dans l'Egypt Pharaunique. Pp. 9-42 in *Les Communautes Rurales= Rural Communities,* Recueils de la Société Jean Bodin pour l'histoire comparative des institutions, ed. J. Gilissen. Paris: Dessain et Torla.

Voeoebus, A.
1977 *An Unknown Recension of the Syro-Roman Lawbook.* Stockholm: ETSE.
1982 *The Syro-Roman Lawbook.* Stockholm: ETSE.

Waddington, W.H.
1870 *Inscriptions grecques et latines de la Syrie.* Paris.

Waston. A.
1983 Agriculture and Law in Rome. Pp. 403-409 in *Les Communautes Rurales, Rural Communities,* ed. J. Gilissen. Paris.

Whittaker, C.R.
1976 Agri Deserti. In *Studies in Roman Property,* ed. M. Finely. Cambridge: Cambridge University Press.

Zacharias Rhetor (or, of Mitylene)
1899 *Historia Ecclesiastica.* ed. tr. Brooks, Corpus scriptorum Christianorum orientalium (Scr. Syr.) 3, 25. Paris: E Typographeo Reipublicae.

Dogs in Ancient Rural Jewish Society

JOSHUA SCHWARTZ

The Martin (Szusz) Department
of Land of Israel Studies
and Archaeology
Bar Ilan University
Ramat Gan, 52900
Israel
schwaj@mail.biu.ac.il

This article examines the history of dogs in the Land of Israel in the rural sphere of ancient Jewish society during the Mishnah and Talmud periods. Particular attention is paid to the sheep or herding dog, the hunting dog, the guard dog and the kufri-dog. While these dogs may have played a less significant role in Jewish society than in the Roman or Persian worlds, it seems that they did have a more important role than what has previously been assumed.

Introduction

This article examines one aspect of the history of dogs in ancient Jewish society, i.e., their place in rural Jewish society in the Second Temple, Mishnah and Talmud periods.[1] The study for the most part deals with the rural society of the Land of Israel, but at times of necessity examines traditions pertaining to these animals in (rural) Babylonia. While our comments are mainly limited to Jewish society, it is impossible, of course, to divorce the attitudes to these animals, whether in rural or urban Jewish society, from those prevailing in the non-Jewish world.

The Rural Dog in Biblical Society

The Near Eastern canine was domesticated at some time during the Epi-Paleolithic Era (15,000-8,000 B.C.E.) and immediately was assigned such tasks as "guard dog" or "hunting dog" (Davis 1987:127), the latter certainly an activity related to the rural sphere and the former also possible there. Despite the potential functions that the dog could have and, indeed, often did fulfill, the attitude of the Bible to the canine was negative (Thomas 1960; Goodfriend 1995; Borowski 1998:133-140). This may have been to some extent due to the fact that the domestication process at the time was far from complete, and dogs are often described in contemporary literature as "settlement predators," regarding both city and village, eating human carcasses or licking human blood (1 Kings 14:11; 16:4; 21:19, 24; 22:38; 2 Kings 9:35-36). The partly domesticated canine is described as feasting on human carrion in settled areas while the birds of the field enjoy similar repasts in the unsettled ones. The semi-wild (or semi-domesticated) dogs would wander the streets of a settlement at night, both city and village, barking, whining, scrounging for food (Ps 59:7, 15), returning to their own vomit (Prov 26:11), and attacking whether provoked or unprovoked (Ps 22:17, 21; Prov 26:17). This may well explain in part the negative attitude in Biblical literature to dogs, regardless of which area they inhabited.

The sense of disgust described in biblical literature seems to reflect the situation in most of the ancient Near East, apart from Egypt, despite the various important functions fulfilled by the dog, even of a cultic or medicinal nature (Collins 1990). Some of these animals may have had their purposes, but they were certainly not man's best friend, whether in city, town or village.

Greeks, Romans and Persians

The Greeks also used dogs for tasks associated with the rural sphere, such as hunting and herding, and as guard dogs in both town and country (Beck 1975: 49). Dogs were also considered to have healing powers, and thus were often left to wander freely in Temples of Asclepius, which were located also in rural regions (Serpell 1986: 175-76). What made the difference in contrast to the attitude described above that was prevalent in the ancient Near East, was that dogs were popular pets in Greek society, among male and female, young and old, rich and poor and would often accompany their masters during the day (Schwartz 2000). This was probably more of an urban than a rural phenomenon, but even dogs fulfilling tasks in rural society, such as sheep dogs, could well have become at least part-time pets.

The only difference between Greek and Roman society in relationships to dogs was that all the above-mentioned phenomena in Greek society were much more pronounced in

Roman society, especially the personal attachment between dog and master (Toynbee 1973:102-24). This was not necessarily an urban as opposed to rural phenomenon, although, obviously, the more socio-economically well off could better afford to pamper their dogs, whether in town or village.

Dogs were also extremely popular in the Persian world, and, indeed, nowhere in the Near East was the dog more revered, occupying a place of the highest respect in the Zoroastrian religion (Wapnish and Hesse 1993: 71-2). There was a good deal of "canine literature," although little is extant, but what remains mentions the types of dogs popular in the Persian world, particularly hunting and sheep dogs (Darmesteter 1895: 160). It was forbidden to harm or improperly feed dogs, and those who did so or mistreated them in any other way were to be severely punished. This reverence was undoubtedly connected to the importance of their functions, and thus even guard or house dogs had a significant place because they watched for and protected against thief and wolf, and the spiny hedgehog -- considered a canine by the Persians-- also rid houses of poisonous snakes (Darmesteter 1895: 156, 160). None of this, of course, prevented the dog from also being a pet, although this was unlikely in the case of the hedgehog. Moreover, as in the Greco-Roman world, almost all the above could also pertain to the rural sphere, and certainly applied regarding hunting and sheep dogs.

The Dog in Jewish Rural Society

The following examines some of the more important roles of the canine in Jewish rural society, dealing first with aspects that are clearly related to rural society and then with those that may also, but not necessarily, relate to this society. When necessary, the Jewish rural motifs are examined in light of the background of the ancient Near East and of the Greco-Roman and Persian worlds.

Sheep Dog - Herding Dog

The main task of the sheep or herding dog was to guard sheep or cattle against predators, whether human or animal, and to round them up and bring them to the shepherd if they had strayed. Eventually, such dogs might also have served to "patrol the boundary," i.e. keep the livestock out of a neighboring field, or as drovers when long distances had to be traversed, but these last two functions were not very prevalent in ancient times.

Even in the ancient Near East, which, as noted above, was not overly enamored of the canine, the sheep dog, because of its importance, was often treated with a degree of respect. Hittite law (Pritchard 1955:193, No. 87) stated that anyone who struck a herdsman's dog and killed it, had to pay the herdsman twenty shekels of silver and pledge his estate as security for payment. This price may have been up to twenty times that of a single sheep, showing the value of a trained sheep dog (Goodfriend 1995: 392). This is also clear from the fact that one who killed a regular dog had to pay only anyone shekel (Pritchard 1955:193, No. 89).

The Bible also mentions sheep dogs and, although their importance cannot but have been recognized, biblical tradition still finds it difficult to break away from its underlying disdain for the canine. Thus sheep dogs (כלבי צאני) are mentioned in Job 30:1, but as a way of debasing those whom Job wishes to insult, in that he would not let those mocking him now even have handled his sheep dogs in the past. Isaiah 56:9-12, also as an insult, attacks leaders who are compared to shepherds and sheep dogs, pointing out that sheep dogs that cannot bark to warn of imminent danger or who are greedy and rapacious are of absolutely no value. The underlying negative attitude of the Bible to dogs, in spite of the importance of the herding dog, can also be seen in Exodus 22:30, which states that a carcass found in the field should not be eaten by the people of Israel, who are holy, but rather thrown to the dogs. These dogs have been interpreted as herding dogs (Goodfriend 1995: 391-92), and although this statement implies that they are being provided with food, a diet of carrion or refuse would hardly reflect an exalted position.[2]

The attitude to the herding dog, in contrast, was extremely positive in the Greco-Roman world. Horace describes them as the "shepherd's friendly force," Virgil emphasizes that they should be cared for and fed well (Toynbee 1973: 106-7) and Columella suggests a diet of barley flour with whey or bread made with bread-wheat mixed with the lukewarm liquid of boiled beans (*De re rustica*, vii.12.10). Columella provides a good deal of information regarding herding dogs. He recommends giving them short names so that they would respond quickly when called (*De re rustica,* vii.12.13). They should be white, so as not to cause confusion when driving off a wild beast in the night or early morning (*De re rustica*, vii.12.3), and they should be lean and vigorous enough to fulfill all their tasks (*De re rustica*, vii.12.9). Columella indeed sings the praises of the sheep dog: "For what human being more clearly or so vociferously gives warning of the presence of a wild beast or a thief as does the dog by its barking? What servant is more attached to his master than is a [sheep] dog?" (*De re rustica*, vii.12.1)

The same positive attitude to the herding dog was also prevalent in the Persian world. Thus, the Zoroastrian *Vendidad* states that the role of the shepherd's dog is that "he comes and goes a *yugyesti* [a distance of 16 *hathras* = 16,000 paces] round the fold, watching for the thief and wolf" (Fargard 13 in Darmesteter 1895: 160). Giving bad food to such a dog is, according to this literature, the equivalent of giving such food to an invited guest in one's house, and the penalty is 400 stripes (Darmesteter: 160-1). Another tradition maintains that indeed not a single head of cattle would remain in existence were it not for the shepherd's dog (Saddar 31, Darmesteter 1895: 168, n.1).

The "Jewish" Sheep Dog in the Talmudic Period

The above-mentioned positive sentiments in the non-Jewish world may not have been as pronounced in Jewish society, but the overall attitude to these types of dogs must have been similar, and for good reason: when grazing sheep or cattle, the herding dog was an absolute necessity. Indeed, Josephus discusses receiving payment "for the mating of a dog, whether hound of the chase or guardian of the flocks" (*Ant.* 4.206; cf. Mishnah Temurah 6:3 and parallels). Clearly, these dogs -- the hunting dog and the sheep dog -- are not "mongrels",[3] but rather, the most important types of dogs that exist, at least according to Josephus.

Ancient Jewish sources mention three kinds of sheep grazing (Safrai 1994: 165): 1. in the desert throughout the year; 2. in the desert during the summer; 3. near a settlement. It is difficult to determine from the few sources available to which of each of the three kinds of grazing the herding-dog traditions relate, but this is not of great importance, because it is likely that a sheep or herding dog was necessary for any kind of grazing, although if a farmer had only one or two sheep, he may have been able to make do without a dog. He may have kept a dog nevertheless, however, to serve in other capacities, such as guard-dogs, as is discussed below. In any case, if a farmer handed over his sheep to a professional shepherd to be grazed (Safrai 1994: 167), there is no doubt that the shepherd would have had dogs at his disposal.

While it may be argued that it was not necessary to maintain sheep dogs for very small flocks or individual sheep, in cases of many or large flocks, it was undoubtedly necessary to use dogs. Thus, we find the following tradition in Genesis Rabbah 73:11 (pp. 855-856, ed. T. Albeck; with variation in Tanhuma, ed. Buber, Va-Yeze 24, pp. 81a-b), in a description of the vast property owned by Jacob when he returned to the Land of Israel:

> Rabbi Abba bar Kahana said: The Patriarch Jacob had one hundred and two myriads, seven thousand and twenty droves. R. Levi said: He had sixty myriads of dogs. The Rabbis said: one hundred and twenty myriads. [*var. lec.* from Tanhuma: And how many sheep dogs did he have?...]. Yet they do not disagree, the number of sixty myriads being taken on the view that there was one dog for each drove, while one hundred and twenty myriads is of the view that there were two dogs for each drove (translation, ed. Soncino, p.675).

The number of droves or herds is clearly fantastic and so, of course, is the resultant number of dogs, and what R. Abba is actually saying is that Jacob was a very rich man. What is more interesting for our purposes, however, is whether there were one or two dogs with each herd. Under normal conditions, an average herd may have been small, numbering between two and five sheep. In such a case, one dog, serving more as a guard than a drover, would have sufficed. The tradition, however, theoretically describes livestock being driven over long distances. The herds would have been larger and the dogs would have had to function both as drovers and as guards. In this case, two dogs per herd or flock, driving the sheep and guarding them, would have been more appropriate.

Rabbinic literature reflects the reality of two dogs per flock. Thus, a tradition in Sifre Numbers 157 (p. 209, ed. Horovitz; variations in Babylonian Talmud Sanhedrin 105a; Tanhuma-Buber Balak 4 [p. 67b]; Tanhuma Balak 3 ; Numbers Rabbah 20:2) states:

> To what may this be compared? Two dogs that were in a flock and they were angry with one another. A wolf came to take a lamb from the flock and one of the dogs was attempting to stop him. The other dog said: if I do not go and help him, he [the wolf] will kill him now and then circle around and kill me. They reconciled with one another and fought against the wolf.

Although we do not know the size of the herd, the source explicitly states that there were two dogs. Theoretically, to be effective, the dogs would have to function in harmony, but they are, after all, only dogs and it is conceivable that occasionally they might not even respond to commands of their master (Babylonian Talmud Bechorot 55a). When the flock is threatened, however, they function in concert against the common enemy. In this instance we also see that having two dogs for a flock was not only in order to drive the herd, but also important for guarding it.

To guard the flock, the dog needed also to be somewhat aggressive and willing to pick a fight and quarrel (Columella, *De re rustica,* vii.12.9). This was the case not only regarding repelling wolves, as seen above, but the dogs at times would have had to face jackals (*Ben Sira* 13:20; cf. Beinart 1998: 178-88) or a pack of wild dogs attacking from more than one direction (Mishnah Bava Mezia 7:9). They might even be attacked by the animals they were herding or protecting, if such were large and potentially dangerous, like oxen (Palestinian Talmud Bava Kama 2:6, 3a). This is, of course, another good reason why shepherds would have preferred two dogs per herd, in order to guard the flock from more than one direction. The sheep dogs would also have been required to guard against thieves trying to make off with the livestock (Babylonian Talmud Bava Mezia 93b-94a).

As stated above, in the non-Jewish world, because of the importance of these dogs, they had to be fed well. Rabbinic literature seems to indicate that this also was basically the case in Jewish society and, at the very least, Rabbinic law required that a master feed his dog, whatever its nature or purpose (Babylonian Talmud Shabbat 155b; cf. Tosefta Yom Tov 2:6, 287, ed. Lieberman). As for the food given to sheep dogs, Mishnah Hallah 1:8 states that if shepherds eat "dog's dough," i.e., which would normally be fed to the sheep dogs (cf. Columella's suggestions regarding diet above), religious rules and regulations attached to dough prepared for human consumption, such as *hallah,* would apply also to this dough. Although it might be argued that this implies the possible mean diet of the shepherds rather than the enriched diet of the dogs, the fact that all the religious prescriptions apply would seem to indicate that the latter is the case. It should be born in mind, however, that all things being relative, using barley flour in food for a dog might be considered an enriched diet for the animal (Columella, *De re rustica,* vii.12.10), while for humans it might be edible, but hardly gourmet fare (Avot d'Rabbi Nathan, version A, chapter 20, 71, ed. Schechter), and, indeed shepherds would not eat every kind of dough prepared for dogs (Palestinian Talmud Hallah 1, 9:58a). As for the composition of the dough, the Palestinian sage R. Simeon b. Lakish stated that coarse grain or bran was added to it (Palestinian Talmud, *ibid.*) and the bread prepared from it might look like regular bread (Tosefta Hallah 1:7, 276, ed. Lieberman). When all was said and done, however, the sheep dog could make do with carrion (Babylonian Talmud Bezah 21a) or with a bone (Babylonian Talmud Shabbat 128a; cf. Josephus, Ant. 12. 288 ff.) but a master might even slaughter an animal for a dog (Tosefta Hullin 2:15, 502, ed. Zuckermandel).

Although the herding dog was mainly supposed to protect the flock or herd, it was also known to be concerned about the life of its master. Palestinian Talmud Terumot 8, 7: 46a relates the following tale:

> Shepherds were milking and a snake came and ate from it and the dog was watching. When the shepherds came they wished to eat from it. The dog barked at them and they paid no attention. Finally, the dog ate it and died.

In this instance, the herding dog was willing to sacrifice its own life in order to save the shepherds. The parallel to this story in Pesikta de Rav Kahana, Va-Yehi be-Shalah 1 (175-176, ed. Mandelbaum) adds that the shepherds, in their gratitude, buried the dog and constructed a monument in its memory, which was then known as "the dog's monument." Although Saul Lieberman (1962: 96, n. 91) claimed that this story and those that follow it in Palestinian Talmud Terumot and Pesikta about dogs saving or protecting their masters reflect stereotyped pro-dog motifs of the Greco-Roman world, the very fact that this motif found its way into rabbinic literature would seem to show that it was also believed in the Jewish world that a sheep dog could behave in this worthy manner, and there was probably good reason for this belief, stereotype or not.

Since the herding dog was known for its faithfulness to both flock and shepherd, exceptions were noteworthy and might indeed reflect a topsy-turvy world. Thus, in the Apocalypse of Animals in the Book of Enoch, in which animals are used to tell the history of mankind and Israel from Adam until the coming of the Messiah, the dogs join forces with foxes and wild boars and attack the sheep instead of protecting them (89:42-49). The sheep have been identified with Israel, the dogs with the Philistines, the foxes possibly with the Ammonites or Amelekites (Black 1985: 267) and the wild boar with the Edomites. Although the animal symbolism is far from clear, the large dog cemetery found in Persian period Ashkelon, formerly a Philistine city, and related to the Phoenician population at the site (Wapnish and Hesse 1993; Stager 1991a; 1991b; Halpern 2000) might help to explain the choice of this animal to symbolize the Philistines. Be that as it may, what is relevant to our discussion is that in this case, the dog acted as a predator instead of protector (cf. Mishnah Bava Mezia 7:9) and in any event, dogs would not usually eat sheep (Babylonian Talmud Bava Kama 15b).

The sources discussed thus far most probably reflect grazing or herding by a professional shepherd relatively far from the home settlement of the herd, but animals may have been also grazed in the vicinity of settlements. These animals, sheep or goats, would have been kept in pens within the dwelling-area which was comprised of a residence and a courtyard for the animals (Safrai 1994: 165). The animals would still have needed protection, although in this case, probably more from thieves than predators, and it is likely that the protection would have been in the form of a dog, whether it was considered a sheep dog or a farmyard guard dog (see below).

Thus Mishnah Bava Kama 2:3 mentions a kid (or goat) and a dog that jumped down from the top of a roof and broke vessels, in which case the owner of the animals would be responsible for the damage. Although there has been a tendency at times to interpret this Mishnah as dealing with a kid *or* a dog, there is no reason not to accept the simple meaning of the text. The dog and the kid were in a settlement and the kid managed to get out of the pen. As it is wont to do, the kid climbed up on to one of the roofs, followed by the dog responsible for its welfare. The courtyard containing the pen was communal, and the animals might, therefore, have caused damage to anyone's utensils or they may have jumped into a different courtyard altogether. It seems clear, though, that neither the goat nor the dog were expected to remain in the pen, and there are instances of their owner even being exempt if they caused damage, for example, if they fell by accident (Tosefta Bava Kama 2:1, 5, ed. Lieberman; Palestinian Talmud Bava Kama 2:5, 3a; Babylonian Talmud Bava Kama 21b) or caused damage when they were jumping up to the roof from the pen or courtyard and down back in to the courtyard (Babylonian Talmud Bava Kama 21b). Needless to say, this did not mean that the goats or sheep did not usually have to be kept confined in the pen (cf. Mishnah Bava Kama 6:1).

The Hunting Dog

Dogs have long been associated with hunting. There were hunting dogs of various types and sizes in the ancient Near East (Bilik and Beinart: 1962: 110-11) and the same applies to the Graeco-Roman world (Toynbee 1973: 102-6). The size and shape of these particular dogs usually related to the prey being hunted and to the particular task of the dog in the chase and its role in assisting the hunter in either capturing or killing the beast. It is no wonder, then, that hunting dogs were apparently highly trained (Toynbee 1973:106). It goes without saying, of course, that these dogs would be found especially in societies that engaged in hunting or held it in esteem.

As for Jewish society, Josephus, as mentioned above, writes of the possibility of receiving payment for the mating of two types of dogs, the guardian of the flock (discussed above) and the hunting dog or dog of the chase (*Ant.* 4.206). Although this might imply that there were many hunting dogs in Palestine, and consequently much hunting, the sources do not seem to bear this out. Indeed, there is only one other source from Second Temple, Mishnah and Talmud periods that appears to refer to a hunting dog.

Chapter 2 of the Testament of Judah tells of Judah's prowess as a hunter. He used to race hinds to prepare as a meal for his father. He chased roes, slew a lion, fought bears, raced the wild boar and "rent like a dog" those wild beasts that turned against him (*Testament of Judah* 2:2-5). The account continues that in Hebron, a leopard pounced upon Judah's dog and, to save it, Judah grabbed the leopard and hurled it all the way to the district of Gaza (*Testament of Judah* 2:6). It is difficult to determine exactly what Judah is supposedly doing in this account, and whether he was hunting for hunting's sake or simply dealing with dangerous animals he happened to encounter. Indeed, it is not even clear whether he initially hunted the hind he prepared as a meal for his father. It should also be born in mind that these wild animals could have been encountered by a shepherd, and the dog, therefore, could have even been a sheep dog. In any case, Judah is described as being accompanied by his dog when he engaged in these exploits, and if he indeed was hunting, then it can be assumed that the dog was a hunting dog, although, as we stated, it could have been a sheep dog or merely a companion (Schwartz 2000).

In any event, it is difficult to ascertain what Josephus had in mind when he mentioned dogs of the chase. Perhaps Jews raised them for sale to non-Jews (cf. Mishnah Avodah Zarah 1:7).

There are no references at all to hunting dogs in Talmudic literature (Lewysohn 1858: 89, but cf. the comments of Rashi on Babylonian Talmud Bava Kama 80a), perhaps reflecting the continued limited nature of Jewish participation in this occupation or pastime (Mainzer 1909; Schwartz 1997 [on fowling]). This might also explain why there are so few chase scenes in Jewish art, in general, and including dogs, in particular, as opposed to the popularity of this motif in non-Jewish art.

Thus, to cite a few examples from Palestine, a stag and a hound are depicted on the mosaic floor of a Roman villa (which subsequently was turned into a Christian chapel) in Beth Guvrin (Avi-Yonah 1981: 293). A mosaic in a room on Mt. Zion belonging to the Augustine Fathers of the Assumption depicts a dog chasing a gazelle (Avi-Yonah 1981:317). The mosaic floor of the church at Kissufim near Gaza portrays a hound chasing a hare and antelope (Hachlili 1988: 339). The floor of the large Roman period building recently excavated in Lod, which probably served a public function, depicts a hare eating grapes and next to it a hunting dog wearing some kind of special harness (Avisar 1997: 50). A mid-third century CE mosaic floor in Nablus depicts a boar being chased by a hunting dog wearing the same kind of harness (Dauphin 1979: 15).

Hunting dogs do appear, however, in a few Jewish material culture remains, for example, on a Jewish lamp mentioned by Goodenough that depicts a dog chasing a hare (1953/68: I, 142).[4] The motif also appears in Jewish mosaics, for instance, in the floor of the synagogue at Beth Shean, that includes a chase scene with a dog pursuing a hare and that at Gaza, with dogs chasing a deer (Hachlili 1988: 339). Although these dogs are clearly hunting, it is difficult to ascertain whether they are indeed "hunting dogs" or rather wild dogs or some other type of dog engaging in "unofficial" hunting activity.

The Guard Dog: Yard and House

Another important task of the dog was to serve as a guard dog (Toynbee 1973: 107-8; Darmesteter 1895: 168). Technically, this function was not necessarily connected to the rural sphere, since a house-guarding dog could also protect a residence in a city. However, since Columella specifically connects the watch-dog functions of house-guard and farmyard dogs (*De re rustica* vii, 12. 2-3) with the latter definitely belonging in the rural sector, the guard dog is included in our discussion, with reference to specifically rural aspects.

The guard dog, according to Columella, should be of ample bulk, with a loud sonorous bark, and thus able to frighten off the potential malefactor by both sound and sight. Unlike the sheep dog, which should be white to distinguish it from animal predators, this dog should be completely black, increasing the fright factor in warding off human intruders and providing for ample camouflage in the nighttime. These dogs must be vigilant, cautious and not given to wandering. It is important that they not be too aggressive, since they might then also attack members of the household. They need not be overly agile, since even farm dogs do not need to patrol that much of a perimeter. Their first duty is not to allow themselves to be attacked, and their second to defend themselves and their territory if they indeed are attacked (*De re rustica*, vii. 12. 3-7). Although this description theoretically refers to the "Roman" guard dog, it is likely that as in the case of much of the material relating to canines, it is fairly universal. With this in mind, we can now turn to the Jewish material.

The inherent value of the dog for security was well known. Thus, a Babylonian tradition states that one should not live in a settlement in which a horse does not neigh or a dog does not bark (Babylonian Talmud Pesahim 113a). Although the source may theoretically refer to any type of dog that barks when confronted by an intruder, it seems unlikely that it refers to a wild dog or scavenger, since these would hardly distinguish between friend and foe or between intruder and resident. It seems more likely that the source alludes to the run-of-the-mill guard dog in the rural sector and that this would be the case in the Land of Israel as well as in Babylonia.[5]

Needless to say, guard dogs had more specialized functions. A midrashic tradition recounts a story of a dog guarding a vineyard (Pesikta de Rab Kahana Zachor 3:9, 47 ed. Mandlebaum; Tanhuma, Ki Teze 12:21a, ed. Buber; Pesikta Rabbati Zachor 12:52a [orchard instead of vineyard]; Pirkei d'Rabbi Eliezer 44 [orchard]):

> R. Levi said: Like a highwayman he came upon you from the wayside. In this connection is told a parable of a king who had a vineyard which he enclosed with a fence and in which he put a dog who was a biter. The king said: If any one should come and breach the fence, the dog will bite him. When the king's own son came and breached the fence, the dog bit him....(translation, 51, ed. Braude and Kapstein)

The story, unfortunately, leaves a good deal unexplained. What is clear is that the guard dog in question was certainly aggressive enough to fulfill its duty and that it was not a house dog or pet transferred to temporary guard duty because then it would have known the king's son. From the source, however, it is difficult to determine whether guard dogs were used by other social strata for the protection of their property. Perhaps only the upper classes or "kings" used them. Also, it is difficult to ascertain whether the dog would have been effective in a situation without a fence, as in the case of most vineyards. In other words, could the dog have patrolled the perimeter?

A similar tradition appears in Kohelet Rabbah 5:10 (cf. Ruth Rabbah 3:2) which relates that King David used to like to spend part of the Sabbath in an orchard behind his house, where he ultimately met the Angel of Death. Upon discovering the death of his father, Solomon informed the sages that "father is dead and lies in the sun and the dogs of father's house are hungry." There are two possible explanations for the presence of these dogs. They were David's companions or pets that accompanied him from his house. In this case, though, one would not imagine that they would be described as ready to cannibalize him. The more probable alternative is that they were probably the guard dogs of the orchard, keeping out intruders and protecting both the fruit and its owner. These dogs might not have been considered that familiar with David and, if hungry enough, might have been described as willing to eat him. The rabbis instructed Solomon to throw the dogs an animal carcass to assuage their hunger. This, too, hardly seems to suggest house-dog companions.

In any event, it should be pointed out that the few vineyard or orchard guard-dog traditions relate to kings, which might indicate that, indeed, only the upper classes had real guard dogs to protect their gardens, orchards or vineyards. Everyone else probably made do, if at all, with whatever type of dog that was available.

As stated above, guard dogs were also of the farmyard or courtyard variety and their task was to guard livestock and property. Some dogs were particularly trained for these tasks and others probably just behaved like dogs, i.e. raising a ruckus when an intruder entered their territory. Thus, Mishnah Bava Kama 5:3 relates that if someone brought an ox into a courtyard without the permission of the owner and if that ox was bitten by the master's dog, the owner of the dog (and courtyard) was exempt, since, after all, the dog was only doing its duty. If, however, permission had been granted to enter, the owner of the courtyard was responsible for the damage caused by his dog, since it should have been restrained. Rabbi Judah the Prince added that even in this case, the owner of the courtyard was exempt, unless he had specifically indicated that he was responsible for the animal being brought in, a view that clearly gives tremendous latitude to the farmyard dog.

The Tosefta (Bava Kama 5:13, 20, ed. Lieberman) offers a different explanation. If an animal came into a courtyard without permission and dies as a result of being gored by the ox of the householder or bitten by his dog, then the animal that caused the injury is stoned, but the householder does not have to pay the ransom to the owner of the dead animal, since it entered without permission. The biblical ruling regarding a ransom in such a case (Ex. 21: 29-30) relates to an ox that gores and it would seem likely that this is the case in the Tosefta and that the ransom has nothing to do with the dog. The entire Chapter 5 of Tosefta Bava Kama, for that matter, deals with damage inflicted by oxen or cattle, strengthening the assertion that the stoning or ransom would pertain only to the ox and not to the dog. It is hard to imagine that the courtyard dog, however well trained (or untrained), could actually kill the intruding beast, although, as will be seen below, damage could certainly be inflicted. In any case, the dog is not listed among those animals, such as a lion, bear, snake etc., that might automatically be considered dangerous (Mishnah Bava Kama 1:4; Palestinian Talmud Bava Kama 1:5, 2c).

An additional *halachah* in the Tosefta (Bava Kama 6:27, 27) states that if a laborer enters a courtyard to receive his wages, even with permission, and is gored by the householder's ox or bitten by his dog, the householder is still exempt. According to Babylonian Talmud Bava Kama 33a, this refers to an instance in which it is not clear what the householder really meant, that is whether his affirmative reply meant that the laborer should actually enter the courtyard or that the householder would soon shortly bring his wages to the entrance of the courtyard, and that the laborer should not enter, perhaps because of the danger. Be all that as it may, it is clear that the dog might attack an intruder, whether human or animal, and it is likely that this indeed was its job and, therefore, the *halachah* seems to be fairly liberal regarding the dog in such cases.

An owner also might be exempt from damage caused by his dog if it occurred during the course of what might be construed as guard duty. However, damage caused as a result of "misbehavior," such as eating a cake being baked in the communal courtyard and belonging to someone else, or causing a fire by taking the cake together with a live spark and igniting something had to be covered by the owner, whether in part, in the case of the fire damage, or entirely in the case of eating the cake (Mishnah Bava Kama 2:3).

The advantages of the guard dog often had to be weighed against possible damages, and there was always an inherent fear, possibly drawing on the biblical period anti-dog ethos, of the dog's potential violence. As seen above, aggressive behavior was often a required characteristic in certain kinds of dogs, but this could get out of hand and it was necessary at times to take precautions. Thus, Mishnah Bava Kama 7:7 states that dogs should not be raised unless they are chained, and one extreme view, that of R. Shimon b. Eleazar, states that dogs should not be raised in a settlement at all because of the inherent danger (Tosefta Bava Kama 5:17, 40-41, ed. Lieberman). Dogs, probably even if they served as courtyard guard dogs, should not be left to wander freely around the yard. These restrictions, though, were hardly unusual, and the Persian tradition, for instance, mentions collars and muzzles for house or courtyard dogs (Darmesteter 1895: 164).

The restrictions could at times be removed. Thus, Babylonian Talmud Bava Kama 83a adds that in border towns the dogs should be chained during the day but should be let free at night. The tradition in the Babylonian Talmud here is apparently adopting the view of R. Shimon b. Eleazar, who allowed dogs to be raised only in such border towns, provided they were chained during the day, although they could be let free at night. In this case, they clearly serve as guard dogs, warning of intruders and perhaps even attacking them. The residents of such towns would either be at home in the evening or perhaps known to the local dogs and therefore left alone.

In general, however, Talmudic tradition, apparently recognizing the contradiction between the advantages of the dog and its potential violence, differentiated between the regular dog and the "evil dog." Thus, Palestinian Talmud Bava Kama 7:7, 6d, commenting on Mishnah (Bava Kama 7:7), adds that whoever raises an "evil dog" prevents kindness from his fellow (or from his own house, cf. Babylonian Talmud Shabbat 63a-b). The dog did require, of course, some dog-like aggressive characteristics, but it should not be vicious and, in any case, steps should be taken to control it.

This, however, did not always help. Pregnant women seemed especially prone to miscarriage after being frightened by dogs. One particularly vehement Babylonian anti-dog tradition relates that the barking of a dog caused a woman to miscarry, bringing the number of souls of Israel below that necessary for the continued presence of the *Shekhinah* among them, and the dog, therefore, was actually responsible for the departure of the divine presence (Babylonian Talmud Bava Kama 83a). Another Babylonian tradition relates that a woman went to bake at a neighbor's house and was frightened by the barking of a dog. The master reassured her that the dog's teeth and nails had been removed, but it was too late and the woman had already miscarried (Babylonian Talmud Bava Kama 83a and BT Shabbat 63a-b). These traditions clearly reflect the underlying ambivalent attitude towards the dog. Even when precautions had been taken, damage could be caused, and it was likely that this applied whether in Babylonia or the Land of Israel, in town or country.

Kufri

The last type of dog we shall discuss in relation to the rural sphere may not even have been a dog. Mishnah Kilayim 1:6 states that the union of the wolf and the dog, and the *kufri*-dog and the fox are considered to be *kilayim* or forbidden, and they should not be mated, despite their resemblance to one another. Palestinian Talmud Kilayim (1:7, 27a) adds that this implies that the dog and the *kufri*-dog could be mated, since the Mishnah did not deal with these two together. This tradition in the Palestinian Talmud also adds that R. Meir forbade the union of these two, because he was of the opinion that a dog was to be considered a domesticated animal, while the *kufri* was a wild animal. The rabbis, however, considered both types of dogs to be wild animals (cf. Tosefta Kilayim 5:7-8, 223). Be that as it may, there was some difference between the two animals and the very fact that *kilayim* was introduced shows that this difference was not minor, even according to the rabbis who played them down.

Other traditions seem to imply that there were indeed differences between the *kufri*-dog and the dog. Thus, Tosefta Bava Kama 8:17 (40-41) mentions a tradition of R. Shimon b. Eleazar which states that although one is not allowed to raise dogs in a settlement, one can raise *kufri*-dogs, martens, weasels, cats, and monkeys, which clean one's house. The cleaning of the house relates to ridding it of vermin, for which cats, for instance, were well known (Schwartz 2000). Although R. Shimon's views on dogs were quite strict, as seen above, he permits the raising and maintaining of what was basically considered a wild animal (Tosefta Kilayim 5:7-8), under much less control than he demanded for dogs, which he allowed to be raised only in border settlements. R. Shimon also allowed these animals, including the *kufri*-dog, to be sold to non-Jews (Tosefta Avodah Zarah 2:3, 462), in spite of the fact that such sales of large or wild animals were not usually allowed because of the damage they might do to others if owned by non-Jews (Tosefta Avodah Zarah 2:2, 462). Indeed, Tosefta Sheviit 5:9 prohibits the sale or even hiring out of these animals to non-Jews. Apparently, in R. Shimon's view, their role in ridding houses of vermin was so important that other strictures could be relaxed, whether in relation to their status as dogs or wild animals.

Identifying the *kufri*-dog is problematic. Some scholars or commentators (e.g., Krauss 1911 [1966], II: 121; Albeck 1958: 102) understood *kufri* as related to *kefar* (village) and explained the phrase as village dogs or dogs belonging to villagers. This would certainly be a rural sector dog. It is difficult to understand, however, how these "village" dogs could have been so different from others that the issue of *kilayim* might be raised in any form or fashion. Saul Lieberman explained the *kufri*-dog as either a small, non-dangerous dog (Lieberman 1955/88, II: 552-553, n. 27)-- not dangerous to humans, but hopefully more so to vermin-- or as a black dog typically found today among rural Arab populations (1955/88, II: 652). Neither of these alternatives, however, is sufficient to explain why of all the varieties of dogs, these should be considered different from "regular" dogs in terms of *kilayim*.

Others scholars maintain that the *kufri* was not really a dog. Feliks (1967: 123) interpreted *kufri* as a type of jackal, but this animal would hardly have been allowed in a house under any type of circumstances (cf. Beinart 1998: 177-188). Lewysohn (1858: 88) claimed that the best hint regarding the identification of the animal was found in the Mishnah itself, which states that the *kufri*-dog should not be mated with a fox. Lewysohn maintained, therefore, that the *kufri* was the progeny of a dog and a vixen. It is also hard to imagine, however, what this animal would be doing in a house.

Despite the problems with these identifications, it seems that both scholars were on the right track and that the *kufri*-dog was not in fact a dog. The Persian canine literature, as cited above, considered the "dog with the prickly back, with a long thin muzzle" (Fargard XIII, Darmesteter 1895: 155-56), i.e., the spiny hedgehog, to be of great importance because it rid houses of vermin such as mice and insects and even snakes. While this animal was well known and important in the Persian world in particular, it was also known in the Roman world (Varro, *Menippearum* 490) and this might be the *kufri*-dog referred to by the rabbis. The hedgehog, after all, is indigenous to Asia and some have large pointed ears, similar to those of dogs, which might account for the confusion (Mendelsohn and Yom Tov 1988:48-51). Also, although foxes, mentioned as forbidden mates for the *kufri* in Mishnah Kilayim 1:6 are actually of the canine family, like the hedgehog they are small, and establish dens in burrows, and like the fox have elongated narrow muzzles and short legs, and when they sleep with their tails curled around them, they might even resemble the hedgehog rolling itself into a ball to evade predators (Mendelsohn and Yom Tov 1988: 193-96). They also feed on some of the same vermin as hedgehogs. This might have led to the mistaken conclusion that they could possibly be suitable mates.

Conclusion

As stated at the beginning of this article, the sources included above represent only a small selection of the dog traditions in ancient Jewish society and deal, for the most part, with those aspects relevant to rural life. Dogs may not have been the most important or necessary component of that society, but in light of the material examined above, it would appear that they played a more significant role in ancient rural Jewish society than we may have imagined.

Notes

1 The present study is based on a larger work by the author entitled *Leisure Time Activities in Ancient Jewish Society* and parts of it will appear in a chapter on pets.

2 See also Mechilta Mishpatim 20 (p. 321, ed. Horovitz-Rabin) which hints at the relatively exalted position of these dogs even in situations which, at first glance, may not be evident by stating that these dogs are more important than slaves.

3 Cf. Babylonian Talmud Hagigah 14b on the prohibition against castrating any kind of dog.

4 Goodenough found the lamp in a shop in Beirut, and notes that a menorah is depicted between the hare and dog, hence its interpretation as a Jewish lamp.

5 "Security traditions", however, tend to deal either with human guards or with physical components like walls. See Safrai 1995: 98-122.

REFERENCES

Ancient Source

Ant.
1930 *Josephus, Jewish Antiquities, Books I-IV*. ed. H. St. J. Thackeray. London: William Heineman and New York: G. P. Putnam's Sons.

Avot d'Rabbi Nathan
1887 *Aboth de Rabbi Nathan*. ed. Salomon Schechter. Vienna: Ch. D. Lippe.

Babylonian Talmud
1886 *Babylonian Talmud*. Vilna.

Ben Sira
1987 *The Wisdom of Ben Sira: A New Translation with Notes*. eds. Patrick W. Skehan & Alexander A. di Lella. New York: Doubleday; Anchor Bible, V. 39.

De re rustica
1960-68 *Lucius Junius Moderatus Columella, On Agriculture, I-III*. eds. Harrison Boyd Ash (I), E.S. Forster & E. H. Heffner (II-III). London: Heinemann and Cambridge, MA:Harvard University Press.

Genesis Rabba
1965 *Midrash Bereshit Rabbah*. ed. J. Theodor and Ch. Albeck. Jerusalem: Wahrman.

Mechilta
1970 *Mechilta D'Rabbi Ishmael*. eds. H.S. Horovitz and I.A. Rabin. Jerusalem: Wahrman.

Mishnah
1958 *Shishah Sidrei Mishnah*. ed. Ch. Albeck. Jerusalem and Tel-Aviv: Bialik Institute and Dvir (Hebrew).

Numbers Rabbah
1878 *Numbers Rabbah*. Warsaw (first edition Constantinople 1512).

Palestinian Talmud
1523/24 *Palestinian Talmud*. Venice.

Pesikta Rabbati
1880 *Pesikta Rabbati*. ed. M. Ish-Shalom. Vienna
(1962) (Tel-Aviv: Esther).

Pesikta de Rav Kahana
1962 *Pesikta de Rav Kahana*. ed. B. Mandelbaum. New York: The Jewish Theological Seminary of America.

Pirkei d'Rabbi Eliezer
1973 *Pirkei d'Rabbi Eliezer*. Jerusalem: Eshkol.

Sifre
1966 *Siphre d'Be Rab*. ed. H. S. Horovitz. Jerusalem: Wahrman.

Tanhuma-Buber
1885 *Midrash Tanhuma*. ed. Shelomo Buber. Vilna.

Tanhuma
1520-22 *Tanhuma*. Constantinople. (first ed.)

Testament of Judah
1983 Testaments of the Twelve Patriarchs (Second Century BC). Pp.795-802 in *The Old Testament Pseudepigrapha, Volume 1*, ed. James H. Charlesworth. Garden City, ed. H. C. Kee. NY: Doubleday and Company.

Tosefta
1881 *Tosefta*. ed. M. Zuckermandel. Trier: Lintz
(1970) (Jerusalem: Wahrman).

Tosefta Zeraim
1955 *Tosefta Zeraim*. ed. S. Lieberman. New York: The Jewish Theological Seminary of America.

Tosefta Moed
1962 *Tosefta Moed*. ed. S. Lieberman. New York: The Jewish Theological Seminary of America.

Tosefta Neziqin
1988 *Tosefta Neziqin: Bava Qama-Bava Bathra*. ed. S. Lieberman. New York: The Jewish Theological Seminary of America.

Varro, Menippearum
1972-80 *Satires menippees de Varron*. Edition, traduction et commentaire par Jean-Pierre Cebe. Rome: Ecole Francaise de Rome.

Studies

Albeck, C.
1952 *Mishnah: Seder Zeraim*. Jerusalem and Tel-Aviv: Bialik Institute and Dvir (Hebrew).

Avisar, M.
1997 The Roman Mosaic in Lod. Pp. 49-55 in *New Studies on the Coastal Plain*, ed. Eyal Regev. Ramat-Gan: Land of Israel Studies, Bar-Ilan (Hebrew).

Avi-Yonah, M.
1981 *Art in Ancient Palestine: Selected Studies*. Jerusalem: Magnes.

Beck, F.A.G.
1975 *Album of Greek Education: The Greeks at School and at Play*. Sydney: Cheiron.

Beinart, N.
1998 The Night of the Jackal: Sheep, Pastures and Predators in the Cape. *Past and Present* 158: 172-206.

Bilik, E., and Beinart, H.
1950 Dog. Pp. 110-13 in *Encyclopaedia Biblica*, IV Jerusalem: Mossad Bialik (Hebrew).

Black, M.
1985 *The Book of Enoch or I Enoch: A New English Edition*. Leiden: E. J. Brill.

Borowski, O.
1998 *Every Living Thing: Daily Use of Animals in Ancient Israel*. Walnut Creek: Altamira Press.

Collins, B. J.
1990 The Puppy in Hittite Ritual. *Journal of Cuneiform Studies* 42: 211-26.
Dauphin, C.
1979 A Roman Mosaic Pavement from Nablus. *Israel Exploration Journal* 29: 11-33.
Darmesteter, J.
1895 *The Zend-Avesta, Part I: The Vendidad*, Oxford: The Clarendon Press.
Davis, S. J.
1987 *The Archaeology of Animals*. London: Batsford (pp. 127).
Feliks, Y.
1967 *Mixed Sowing and Breeding: Yerushalmi Kilayim*. Tel-Aviv: Dvir (Hebrew).
Goodenough, E.R.
1953/68 *Jewish Symbols in the Greco-Roman Period.* New York: Pantheon Books.
Goodfriend, E.A.
1995 Could keleb in Deuteronomy 23:19 Actually Refer to a Canine?" Pp. 381-97 in *Pomegranates and Golden Bells: Studies in Biblical, Jewish and Near Eastern Ritual, Law, and Literature in Honor of Jacob Milgrom*, eds. D. P. Wright, D. N. Freedman and A. Hurvitz, Winona Lake, IN: Eisenbrauns.
Hachlili, R.
1988 *Ancient Jewish Art and Archaeology in the Land of Israel.* Leiden: E. J. Brill.
Halpern, B.
2000 The Canine Conundrum of Ashkelon: A Classical Connection? Pp. 133-144 in *The Archaeology of Jordan and Beyond: Essays in Honor of James A. Sauer*. Studies in the Archaeology and History of the Levant 1. Ed. L.E. Stager, J.A. Greene and M.D. Coogan. Winona Lake: Eisenbrauns.
Krauss, S.
1911 *Talmudische Archaeologie*, Leipzig [Hildesheim: (1966) Georg Olm].
Lieberman, S.
1962 *Greek and Hellenism in Jewish Palestine.* Jerusalem: Bialik Institute (Hebrew).
Lieberman, S.
1955/88 *Tosefta Ki-Fshutah.* New York and Jerusalem: The Jewish Theological Seminary of America (Hebrew).
Lewysohn, L.
1858 *Die Zoologie der Talmud.* Frankfurt am Main: self-published.
Mainzer, M.
1909 Jagd, Fischfang und Bienezucht bei den Juden in der tannaeischen Zeit. *Monatsschrift für Geschichte und Wissenschaft des Judentums.*53: 174-89; 303-27; 453-68; 539-62.
Mendelsohn, H.& Yom Tov, Y.
1999 *Plants and Animals of the* Land *of Israel, VII, Mammals*, Tel-Aviv: Ministry of Defense Publishing House and Society for the Protection of Nature (Hebrew).
Pritchard, J. B.
1950-69 *Ancient Near Eastern Texts Relating to the Old Testament.* Princeton: Princeton University Press.
Safrai, Z.
1994 *The Economy of Roman Palestine.* London and New York: Routledge.
1995 *The Jewish Community in the Talmudic Period.* Jerusalem: The Zalman Shazar Center for Jewish History (Hebrew).
Schwartz, J.
1997 'Pigeon Flyers' in Ancient Jewish Society, *Journal of Jewish Studies* 48: 105-19.
2000 Dogs and Cats in Jewish Society in the Second Temple, Mishnah and Talmud Periods. Pp. 25*-34* in *Proceedings of the World Congress of Jewish Studies, Division B, History of the Jewish People.* Jerusalem: World Union of Jewish Studies.
Serpell, J.
1986 *In the Company of Animals: A Study of Human-Animal Relationships.* Oxford: B. Blackwell.
Stager, L. E.
1991a When Canaanites and Philistines Ruled Ashkelon. *Biblical Archaeology Review* 17/2: 24-43.
Stager, L. E.
1991b Why Were Hundreds of Dogs Buried at Ashkelon? *Biblical Archaeology Review* 17/3: 26-42.
Thomas, D. W.
1960 Kelebh 'Dog': Its Origin and Some Usages of It in the Old Testament. *Vetus Testamentum* 10: 410-427.
Toynbee, J.M.C.
1973 *Animals in Roman Life and Art.* Baltimore and London: The Johns Hopkins University Press.
Wapnish, P. & Hesse, B.
1993 Pampered Pooches or Plain Pariahs? The Ashkelon Dog Burials. *Biblical Archaeologist.* 56: 55- 80.

Street Villages and Rural Estate Centers: The Organization of Rural Settlement in the Latin Kingdom of Jerusalem

ADRIAN J. BOAS

The Department of Land of Israel Studies
Haifa University
Mt. Carmel
Haifa, 31905
Israel
badrian@univ.haifa.ac.il

During the twelfth and thirteenth centuries the Frankish leaders and settlers in the Levant became involved in various aspects of rural life, relating to both administration of the indigenous peasantry and direct involvement in agricultural activity. These different facets find expression in archaeological remains, notably rural estate centers, farmhouses and street villages. This paper will present the physical evidence for these activities and how the developed during the Crusader period.

Introduction

The two centuries of Frankish rule known as the Crusader period (1099-1291 C.E.) left a distinct and durable mark on the cities and countryside of the Latin East. However, while in the cities, medieval churches, covered markets and houses illuminate different aspects of urban day-to-day life, the impression still lingers that the impact of this period on the countryside was almost solely the result of military activity, in the form of monumental fortresses built to defend the borders and roads. In fact, this is only one facet of Frankish activity in the countryside of the Kingdom of Jerusalem in the twelfth century -- the period during which (or during most of which) the entire hinterland was under their control.

The castles play a dual role in this picture, both military and non-military. The military role was that by providing security and putting an end to Muslim incursions into Frankish territory, the castles made settlement possible in the countryside. The earliest settlements were villages established in the shadow of the castle walls. The Crusader historian, William of Tyre, described the development of villages around the castles that were built around Fatimid Ascalon in the mid- twelfth century, at Gaza, Betgibelin (Beth Guvrin), Blanche Garde (Tell es-Safi) and Ibelin (Yavneh). Of Blanche Garde he wrote: "Those who dwelt in the surrounding country began to place great reliance on this castle as well as on the other strongholds, and many suburban places grew up around it. Numerous families established themselves there, and tillers of the fields as well. The whole district became much more secure, because the locality was occupied and a more abundant supply of food for the surrounding country was made possible" (William of Tyre 1943: 15,25, p.132).

Other settlements were established more distant from castles and city walls but nonetheless were a direct result of their presence. The many villages and farms established north and west of Jerusalem, for example, were able to survive because these castles in the south put an end to the invasions of Muslim raiders into the Frankish countryside and because, in times of danger, the settlers could always flee to the city for protection.

The second, non-military role of the castles in rural settlement, discussed below, was their function as centers of rural administration.

The Organization of Frankish Rural Activity

The belief that throughout the Crusader period the Frankish population was located almost exclusively in the cities and that agricultural activity remained primarily in the hands of the local non-Frankish population, has been challenged by recent studies (notably Ellenblum 1998), and it is generally accepted today that the Franks were not only rural administrators but also played a certain role as farmers. The question of the extent of their involvement in the latter sphere remains open. Under Frankish rule, the rural population was segregated into separate villages of Syrians (Eastern Christians), Muslims and Franks.[1] Of the estimated 1,200 villages in the Kingdom of Jerusalem, it has been calculated that about 235 (almost 20%) were Frankish settlements (Ellenblum 1998: Map I. See also Prawer and Benvenisti 1970: Part IX, Sheet XII; Prawer 1972: 356 and note 3). If reliable, this represents a fairly substantial number of settlements. However, some of these were certainly very small (the number includes isolated farm-houses and manor houses), and the principal role played by the Franks in the countryside was probably that of administrator rather than settler. As administrators they were mainly involved in keeping the order, deciding disputes between the local and Frankish peasantry, and collecting taxes and dues.

After the conquest of the countryside and the establishment of Frankish rule, the new administration replaced that of the Muslims, which had been based on the *Iqta`* (urban and rural estates, tax revenues or customs dues received in exchange for military and other services rendered to the government in Damascus and Baghdad). Under the Franks, the land was held by the crown, the barons, churches and monasteries, the military orders and smaller landowners. A basic feature in Western feudalism, compulsory service in the lord's demesne, was almost nonexistent. Rather, an annual monetary payment or a portion of the produce was paid to the landowners by the tenants. The annual land tax, based on the Muslim *kharaj*, was known in its latinized form as *carragium*, *terragium* or *terraticum* (terrage). It consisted of a quota of the harvest of arable lands, olive groves and vineyard, ranging from a quarter to half of the produce, but generally a third (Richard 1985:256).[2] Other taxes included a tax on the transport of grain to granaries (*portagium*) and on the use of threshing floors (Prawer 1972:376; Richard 1985:257). Payments were required for pasture rights. Muslim peasants had to pay a tithe or *dime* to the Church (Runciman 1952, Vol. 2:298-99) and the military orders also occasionally collected tithes.[3] The range of produce covered by tithes was broad, including livestock, fodder, reclaimed land (*noval*), first fruits, orchard and garden produce, wood, wool, flax and fish (Riley-Smith 1967:43, 376, 381). The mandatory gift known as *xenia* was a payment made to the clergy on festival days, generally consisting of eggs, fowl, cheese and wood (Tafel and Thomas 1964:371).

There is archaeological evidence of the collection of taxes and tithes in the rural manor houses or estate centers discussed below. These buildings generally included a number of large, vaulted store-rooms, the Near Eastern equivalent of the European tithe barn.

An innovation of the Frankish rural administration was the introduction of the *bannum*, a monopolistic right of the lord to construct mills, ovens and other installations and require the peasants to use them and to pay for their use. Such seigniorial monopolies are recorded in both urban and rural settings: in the estates of the Church of the Holy Sepulchre near Jerusalem (Prawer 1980:134), at Akhzib (Strehlke 1869: No.I), and at Acre, in the agreement between the representatives of the Kingdom of Jerusalem and the Venetian merchants, known as the *Pactum Warmundi* (William of Tyre 1986:12.25). Archaeological evidence for these monopolies has been found in a number of villages and estate centers that have mills and baking installations (e.g. at el-Qubeibeh and Har Hozevim).

Local Settlement and Frankish Administration

In the Crusader period, rural settlements took the form of manor houses, isolated farms (*curtiles*) and villages (*casalia*). The local village in its most complete form included the house of the *rais* (village head), which was often merely a slightly larger house than the others in the village, a church or mosque, the simple dwellings of the villagers, and communal installations, such as cisterns, threshing floors, dovecotes, mills and ovens. The *casalia* differed considerably in size. In the Lordship of Tyre, for example, the population ranged from three to 38 families per village, and in the region of Ascalon, there were roughly 20 families per village (Prawer 1980:160, 167; Tafel and Thomas 1964:398). The *rais* (*raicus* in Latin) was chosen from one of the families (*hamula*) to serve as an intermediary between the Frankish overlord and the peasants (known as *villani* or *rustici*). This was an important position in the rural administration. Aided by a council of elders, the *rais* supervised farming, attended to the collection of taxes and administrated justice. Other members of village administration were the *dragoman* (interpreter) and *scribanus* (scribe). The Court of the *Rais* dealt with disputes between the villagers and decided on fines for offenses against the regulations. The village administration may also have been responsible for the upkeep of and repairs to communal property.

In Frankish villages, the status of the *villani* differed from the local Muslim population and of the *villani* in the West. They were free settlers who could, should they so desire, alienate their land (although the landlord had preemptive rights). As noted above, they were not required to cultivate the lord's demesne, a basic constituent of serfdom in the Western feudal system. Indeed, the seigniorial lord in the Kingdom of Jerusalem often did not retain any land for his own use and the villages were composed entirely of peasant holdings, with the settler devoting his labor to his own land. As a result, the Frankish peasant in the East, rather than struggling against the heavy yoke of feudal life, had a productive and fairly comfortable existence, and it is therefore not surprising that there are reports of Frankish burgesses leaving cities to take up a rural life. They did, of course, have to make certain recompense for the land they received -- i.e., the annual *terragium* mentioned above -- and, like the local *villani*, they were required to use the lord's mills and ovens.

The Frankish Villages

Organization

In addition to being economically dependant on the towns, Frankish settlements, more so perhaps than their Muslim counterparts, depended for security on the town and castles. Raiding parties from across the border posed a greater threat to Christian settlers and particularly to Franks, than to Muslim villagers. Thus, most Frankish rural settlements were located in close proximity to the major towns, specifically around

Jerusalem, Acre and Tyre. Settlements of Frankish *villani* were established either alongside local Christian peasants in existing villages or in new villages. There is documentary evidence of the former, and of the latter, considerably more interesting phenomenon, there is both documentary and archaeological evidence. These new villages were established along the lines of the Western European *villesneuves* (new towns), and were organized, administratively and sometimes physically, in a manner that would facilitate a speedy and efficient parcelling out of the land.

In the Kingdom of Jerusalem, such settlements date to the twelfth century, and it is doubtful that any of them were established after the Franks lost the hinterland following the Battle of Hattin in 1187, or were resettled when they regained parts of it under the terms of the treaties of 1229 and 1240.[4] Both the physical and organizational attributes of these settlements bear a similarity to their Western counterparts. The main physical differences resulted from the use of stone rather than wood in construction. The principal organizational and social difference was that the settlers were freemen rather than serfs.

A useful source of information on Frankish rural settlement activity and the organization of the *villesneuves* takes the form of recorded agreements between landowners and settlers in the villages established in the twelfth century: *Betgibelin* (Beth Guvrin) and at *Casal Imbert* (Achzib) (Prawer 1980:120-26, 140-42). *Casal Imbert* was a royal foundation established in the middle of the twelfth century. According to the terms of the agreement dated to 1153, the settlers received houses rent-free, together with a plot of land for tilling and for a vineyard and garden. Each peasant paid to the king the *terragium*, amounting to a quarter of his crop and two-thirds of the produce from the olive harvest. Olive growing was organized on a communal basis, with individual groves but cooperative harvesting. The king held the *bannum* over certain installations including the bakery and bath-house.

At *Betgibelin*, the Hospitallers founded a settlement of 32 families around their fortress some time before 1164. The settlers received a *charte de peuplement* from the Hospitaller master, which was confirmed in 1168 and again after 1177 (Delaville le Roulx 1894:272-73, no. 399; 350; no. 509). In return for paying the *terragium*, each settler received a plot of land measuring two *carrucae* for cultivation and on which to build a house.[5]

The villagers held additional lands beyond these fields, including olive groves, vineyards and pasturelands, the last including wastelands known as *gastinae*.[6]

Other than the settlements founded by the king or by the military orders, there were also ecclesiastical establishments set up by the canons of the Church of the Holy Sepulchre and by other wealthy monasteries. The Church of the Holy Sepulchre owned 21 villages in the Jerusalem area. One of these, *Magna Mahumeria* (el-Bira, now part of Ramallah), was established in the third decade of the twelfth century (Abel 1926; Pringle 1985). The terms of agreement were not unlike those of the royal and Hospitaller settlements. The villages of the Church of the Holy Sepulchre were primarily involved in grape and olive cultivation and the production of wine and oil for liturgical use, but also, undoubtedly, for sale in the towns and perhaps even for export.

The Archaeology of Planned Villages

The most remarkable innovation of Frankish rural settlement was the adoption of the linear plan village, often referred to as the street- or string-village (Figure 9-1). The planned villages built by the Franks in the twelfth and thirteenth centuries had very little in common with the traditional Near Eastern nucleated village. The typical local village had houses constructed around an open communal area containing public buildings. The other houses were often built against one another to provide a type of defensive outer wall. Such villages developed over an extended period of time and show no evidence of intentional planning. In contrast, the planned village introduced by the Franks consisted of a central, axial road lined on each side by a single row of houses built on long, narrow plots.

The use of the linear plan was a significant innovation in the Near East. Its advantages are clear -- the land could be quickly and evenly parcelled out. The disadvantages are also easily seen -- this type of settlement is extremely difficult to defend. Because of its elongated form, it is difficult to encompass within fortifications and impossible to defend without them. Therefore, the fact that the Franks were able to employ such a design could be seen as proof of the effectiveness of their efforts during the first half of the twelfth century at improving the internal security of the area under their rule. The neutralization of Fatimid Ascalon, the only city that remained in Muslim hands after 1124, was, it would seem, successfully achieved by the construction of the ring of castles around it. It ceased to pose a threat after the middle of the century and, in any event, fell to the Franks in 1153.

The planned settlements built by the Franks do not seem to have been very numerous. At present there is archaeological evidence for only four such villages: *Parva Mahumeria* (el-Qubeibeh), el-Kurum, *Magna Mahumeria* (el-Bira), and a village in *Vallis de Cursu* (Wadi el-Haramiyeh), all located in the area north of Jerusalem (Figure 9-2).[7] Even if there were several more, and there may well have been, it is clear that such planned villages were an exception in the Near Eastern landscape.

On the basis of the information from excavations and surveys, a fairly accurate reconstruction of the typical planned village is possible. A manor house was located at the center of the village (el-Bira, el-Qubeibeh), or on a hill above it (al-Kurum, Wadi el-Haramiyeh). The church was located at one end of the village (el-Bira) or at its center (el-Qubeibeh). There were two rows of houses, barrel-vaulted and solidly constructed of local stone, measuring on average ca. 4x10 m (internally) and probably often with two stories (el-Bira). The houses had thick side walls supporting the vaults, generally over 1, sometimes close to 2 m thick, built with fieldstone with a rubble and mortar fill. They were plastered on the interior. The adjacent houses shared the side walls and the front and back walls often ran the length of several houses (evidence of the setting up of these villages within a short time, as a single construction enterprise).

In the houses with two stories, work and storage areas were located on the ground floor and living quarters on the upper story. On the ground floor of many of these houses, wine and oil presses and other installations were found. Some had fireplaces (el-Kurum, Wadi el-Haramiyeh), a unique feature in this period. Access to the upper story, where there is no archaeological evidence for staircases, was probably by a wooden staircase or a ladder. Some houses also had an undercroft reached by stairs from inside the house or directly from the street. Water was collected from the roofs and fed into cisterns via pipes and channels that ran the length of the streets. The fields were defined by terrace walls extending from the side walls of each house.

Rural Estate Centers

Numerous ruins of towers and manor houses scattered in the agricultural areas throughout the Kingdom of Jerusalem are evidence of the manner in which Frankish landowners or their representatives exercised administrative control over the countryside.

The Role of the Castles in Rural settlement

R.C. Smail noted that the location of castles in the Kingdom of Jerusalem indicates motives other than military for the construction of many of them (Smail 1987:206-7). He was referring to the fact that most twelfth century Crusader castles were built in densely settled agricultural areas, among villages and farms, and not, as one might expect, along the borders or on major routes. The role played by these castles was military only in the sense that in times of danger, citizens could take refuge in them. Many of the castles were of the most basic and simple form, that is, they were fortified towers (keeps), massively constructed rectangular structures of two stories, but quite often without any substantial outworks and completely useless in any offensive role (Pringle 1994a).

It would seem therefore that the main function of these castles was as an early form of rural administrative center or manor house. Perhaps they were constructed in the period when the Franks first got up the courage to come out from behind the city walls and take an active role in settling the countryside, probably around or shortly before the fifth decade of the twelfth century, when the first wave of Frankish rural settlement began.

Let us consider two examples of such buildings: the Red Tower (Burj al-Ahmar) near Tulkarm in the central Sharon plain and the tower at ar-Ram, a village located nine km north of Jerusalem. Burj al-Ahmar, partly excavated in 1983 and published in a detailed report (Pringle 1986), was a two-story tower measuring 11x10.5 m, with two barrel-vaulted halls on the ground floor and six groin-vaulted bays on the upper floor. It was surrounded by a wall, which had no evidence of defensive works, such as salient towers, that would have been necessary for it to function efficiently as an *enceinte*.

The tower at ar-Ram has not been excavated, but has been examined and described by Pringle (1983:165). It measured 12x15 m, with walls nearly 3 m thick. The main entrance was on the east and there was a second doorway to the north. The stairs were incorporated within the thickness of the wall. This tower constituted the first stage of a courtyard building that belongs to the next category discussed in this paper: the manor house. It therefore represents the link between the use of towers for Frankish administrative purposes and the less-defensive hall-houses (houses with a first-floor hall and storage below) and courtyard buildings that came to be the typical estate centers in the Kingdom of Jerusalem

The Manor House and the Farmhouse

A number of rural buildings, generally referred to as manor houses, were to be found throughout the populated areas under Frankish rule in the twelfth and thirteenth centuries. The term "manor house" is not ideal. The Western manorial system with its serfs and demesne lands does not apply in its entirety to the Latin East where there were no serfs and there is almost no record of demesne lands. Nonetheless, this term is frequently applied and, considering that even in the West it was often used in a very loose fashion, it is used in this paper for convenience sake.

As isolated farmhouses are not a typical feature of the local countryside of the Near East in the Middle Ages, the Crusader manor houses are fairly easy to recognize. In addition, their chronology can be established with little difficulty because of the use of well-known Frankish construction techniques and architectural features. The function of these manor houses as centers for the collection of taxes and tithes in the form of farm produce is apparent in the presence of archaeological remains of large vaults that could have served as stables and storage space. The manor house also served as the residence and administrative center of the landowner or his caretaker or representative (*locator* or *raicius*). The ground plans always include a hall-house, occasionally a tower and sometimes both. The hall-house rarely stood alone, although, as the most massive part of the complex, it is sometimes all that remains. It was probably almost always enclosed within walls or incorporated with other vaulted structures.

The typical hall-house was an elongated structure with the ground floor used for storage or stabling and living quarters in a large first floor hall or *solar*, reached by stairs often built within the thickness of the walls. In only a few cases has the upper-story hall survived, but there is often evidence for it in the presence of such a staircase. At *Bethaatap* (Beit `Itab), 17 km west-southwest of Jerusalem, there is a barrel-vault which Pringle has identified as the undercroft of a hall. It measures 13.3x29 m and constitutes the earliest phase of what developed into a characteristic courtyard building (Pringle 1997:26-27). The 1.39 m doorway in the middle of the southern wall has a defensive feature: slit machicolation over the single-leaf door. A staircase within the wall originally led up to the first-floor *solar*.

Another hall-house, Khirbet el-Burj/Khirbet el-Jauz is located above the village of Khirbet el-Kurum in the modern Jerusalem neighbourhood of Ramot Alon (06), south of *Montjoie* (Nebi Samwil). This is an elongated structure measuring 15x18 m (internally). The entrance to the ground floor is on the east. The fieldstone walls are 1.4-2.35m thick. There are four wide windows on the ground floor, an exceptional feature in such a building, suggesting that in this particular case, the ground floor perhaps also served as living quarters. The interior was divided by an internal wall, and there was a cistern beneath the floor and a staircase built into the thickness of the eastern wall.

Together with the hall that served as the living quarters of the landlord (or his representative), most manorhouses also included enclosed courtyards or farmyards, additional barrel-vaulted structures that could have served for storage, stabling and shelters, and various installations, including those that were *bannum*.

These courtyard-buildings took one of two forms: buildings consisting of one or more vaults with a walled enclosure attached to them; or square or rectangular buildings constructed around a central courtyard. Buildings of the former category generally have a single, large, two-story vaulted hall and a number of smaller vaults surrounding the walled courtyard. Their irregular plan reflects their development from the single main vault to a larger complex with several vaults and installations. This perhaps constitutes evidence of the expansion of Frankish rural activity in the second half of the twelfth century.

A farmhouse in the modern Jerusalem neighbourhood of Har Hozevim was completely excavated in 1993-1994 and subsequently destroyed to make way for the construction of a factory (Kletter 1996; May 1997; 2000). The main building was a two-story hall-house, the ground floor of which, measuring 12x19 m, was roofed with a barrel-vault. The main door of the hall was on the east and there was a smaller secondary door in the western wall. As the hall was built on a slope, the higher elevation of the bedrock on the west formed the floor while, on the east, the floor was completed by levelling with sand and gravel and paved with stone flags.

The ground floor had no windows and the only source of natural light would have been through the doors. Despite the poor lighting, this level was used, not only for storage, but also as a work area. In this case, a bakery constituted the main activity. On the eastern side of the room were three vault-shaped ovens, two *tabuns* (clay ovens used for baking) and a stone table. There was also a large plaster-lined cupboard built into the eastern wall for storage. This installation would have been the seigniorial bakery used by the neighboring villagers, who no doubt also brought their wheat there to be milled.

Access to the first-floor hall was via a staircase built in the thickness of the wall on the north side of the main doorway. When it was originally built, this hall-house apparently stood alone. At a later stage, two smaller barrel-vaulted rooms were added, one on each side of the main door, a fieldstone wall was built to enclose the courtyard, and additional vaults and livestock enclosures were built within the courtyard.

The remains of a similar complex, located northwest of Jerusalem at Khirbet el-Lawza, were partially excavated in 1986. This complex was built on three stepped terraces (Ellenblum, Rubin and Solar 1996). It consists of a massive barrel-vault measuring internally 21x5.5 m, probably the first stage of the building which, like the farmhouse at Har Hozevim, was subsequently expanded into a larger complex. Here, too, there was probably an upper-story *solar*. Although no evidence of a staircase was found during the brief excavation, part of a plastered floor apparently from the upper-story level was uncovered (Ellenblum, Rubin and Solar 1996:190). A group of smaller barrel-vaulted buildings and a fieldstone wall enclosed a courtyard measuring ca. 29x40 m. The main gate of the courtyard was on the east, south of the hall. Southeast of the gate was a large, trapezoid-shaped, plastered reservoir that apparently provided water for the agricultural activities of the settlers.

Another remarkable example of this type of manor house is the ruin known as Baldwin's Tower (Burj Bardawil), situated 32 km north of Jerusalem, above the road leading north to Nablus. This complex was originally a castle of the *castrum* type (an enclosure castle with projecting two-story towers and turrets; Pringle 1994:38). As such, it is comparable, albeit on a much larger scale, to the tower at ar-Ram. The castle measured 22 m square. From this nucleus, the manor house developed on the slope to the northeast. A series of barrel-vaults was constructed, forming a fan-shaped enclosure. The largest of these vaults is still known as *el-Babariyya* (an Arabic form of the Frankish term for cow-shed, *bovaria;* See Benvenisti 1982). It is 41 m long, the internal width is 6.9 m and the side walls are 3 m thick. The other vaults enclosing the courtyard are 6.25 m wide (internally), with walls 2.4 m thick. A gateway, 2.5 m wide, was located just to the north of *el-*

Babariyya. The complex may have served as an administrative center for the planned village in the valley below (Wadi el-Haramiyah).

The second type of manor house was a regular, quadrangular building enclosing a central courtyard. The second stage of the manor house at ar-Ram is of this type, with barrel vaults surrounding a central courtyard [Figure 9-3 (Pringle 1983:165-69)]. Beginning as a tower, the complex expanded in stages until it came to consist of four large and a number of smaller barrel vaults. The large vaults were ca. 7 m wide (internally). The now-collapsed vault to the north of the tower, the western range of the complex, was 22 m long. It was the first of the vaults to be constructed. The northern vault was the largest, 45 m long, 7 m wide internally, with two doors in the south. The eastern range was 30 m long and ca. 11 m wide, with two doors and three windows in its western wall. A small, half-vaulted range ran north-south between the eastern and southern ranges. As it partly obscures two windows in the southern wall of the eastern vault, it is clearly of a later date, as is the abutting southern vault range. The southern range was at least 25 m long. There was an outer range to the west of the tower which must be dated somewhat later than the rest of the complex. Also, at the northwest corner, remains of what may have been another, later tower were found. The courtyard enclosed by the vaults covers an area of 30x25 m.

There is a large, rectangular courtyard-building at Jifna, 21 km north of Jerusalem on the road to Nablus (Benvenisti 1970:234, 238-40; Pringle 1997:57). On the east, a monumental fortified gate with a portcullis, gives access to the courtyard, on the north side of which there is a large barrel-vaulted hall of 23.0x7.40 m (internally) and 6 m high. As at Burj Bardawil, this vault is known as *al-Babariyya*, suggesting its function in the Crusader period. It contains an olive press which, judging from its size, (larger than the door), seems to be original. The walls are constructed of fine-quality masonry and are 2.40 m thick.

Perhaps the best known of the manor houses is the well-preserved building of Aqua Bella, located opposite *Fontenoid* (Abu Ghosh) on the road from Jaffa to Jerusalem [Figure 9-4 (Enlart 1928:103-6; Benvenisti 1970:241-45; Pringle 1992; 1993:239-50)]. It is a rectangular building measuring 27x36 m, built into the side of the valley. The northern side of the ground floor was cut into the limestone of the valley slope and the stone that was removed was roughly shaped for the construction of the building. The softer stone used for architectural details such as door jambs, lintels, window frames, stairs and quoins was brought from more distant quarries. These ashlars have the typical Crusader diagonal tooling and frequently display masons' marks (indicators of Frankish construction).[8] The building surrounds a square courtyard measuring 15x15 m, entered from the east. On three sides of the courtyard are barrel-vaulted halls, probably used for stabling and storage. The southern vault contains an olive press. To the south of the courtyard, a stone staircase leads to the upper level where there is a groin-vaulted chapel, an unusual feature in this type of building, but one that could be expected in a building belonging to the Hospitallers (the military order of knights who lived according to a monastic rule). The other vaults on this level served as a residence.

Conclusions

It seems probable that organized Frankish villages of the linear-plan type date to the 40 years between the construction of the castles isolating Fatimid Ascalon in the 1130s and 1140s and the loss of the hinterland in 1187. The apparently more widespread establishment of rural estates and the construction of administrative centers on these estates probably began earlier in the twelfth century. The first seeds of Frankish rural settlement may have been the numerous small castles built in agricultural areas, which served as living quarters for the Frankish administrators. Later came the manor houses, more appropriate for this task, with their spacious living quarters, numerous large vaults for storage of produce and livestock and for various farm activities, and enclosed courtyards. The manor houses were also less defensive in design as a result of improved security within the Kingdom.

The linear-plan villages may have appeared at approximately the same time as the manor houses, around the middle of the twelfth century. The small number of these villages is yet to be explained. It may be that there are many more (two of the known villages were discovered only in the 1990s!). Certainly, the conditions that allowed for the construction of such settlements, particularly a fairly high degree of security in the countryside, did not last very long. However, it could also be maintained that the overwhelmingly greater number of estate centers rather than new Frankish villages, is evidence that the period of Frankish activity in rural areas in the twelfth century was characterized by a predominantly administrative role.

Notes

1 The sources are not all in agreement as to who made up the bulk of the peasants -- Muslim or Eastern Christians. The descriptions of the Frankish chronicler, Ernoul and Ibn Jubayr, a Granadan pilgrim, suggest that there was a Muslim majority (Ernoul 1871:28; Ibn Jubayr 1952:316), whereas the Andalusian, Ibn al-`Arabi, maintained that the Eastern Christians were in control of the countryside (Ibn al-`Arabi discussed in Gil 1992:171). Benjamin Ze'ev Kedar suggests that these observations reflect the situation in different parts of the Kingdom (Kedar 1990:148-49). Joshua Prawer considered the indigenous peasant population to have been predominantly Muslim (Prawer 1969).

2 According to Ibn Jubayr, the Muslim peasants yielded half of their harvest to the Franks and paid a

poll tax of one dinar and five qirat each (Ibn Jubayr 1951:316).

3 Tithes were paid to the Bishop of Margat for villagers, mills and olive presses, gardens and demesne lands (Delaville Le Roulx 1894:595-96, No. 941). Pope Paschal II granted the Hospitallers tithes from demesne lands that they themselves cultivated.

4 In 1229, Frederick II reached an agreement with the Ayyubid sultan which returned Jerusalem and a corridor of land between the Holy City and the coast to the Franks, and in 1240, Richard of Cornwall obtained a substantial extension of land between Jerusalem and the coast, as well as most of the Galilee. However, the internal security achieved in the mid-twelfth century was not regained; Jerusalem and much of the recovered land was lost in 1244, and it is unlikely that there was much, if any, settlement activity in this period.

5 The *carruca*, or ploughland, has been reckoned by Prawer at 35 ha (86.45 acres; Prawer 1980:158) and at a more realistic 3-4 ha (7.4-9.9 acres) by Ellenblum, on the basis that it must have been close to the European *carruca* (Ellenblum 1998:98-99, n. 15). It is estimated that 36-40 bushels of corn were the minimum requirement for a normal family. In the West, a crop of this size necessitated holdings of between five and 10 acres, probably closer to 10 (Bennett 1960:95). In the Latin East, the Frankish *villani* generally received one or two *carrucae*. Archaeological evidence of land divisions is almost non-existent. A single instance in which the fields of a Frankish village can be identified and measured is at el-Kurum, northwest of Jerusalem, where the average field is 8 x 46 m, that is, a mere 368 square m. This is nowhere near Prawer's "official" *carruca*, but it is exactly half of what he calls the "local" *carruca* which appears to be equivalent to the Western *toft* or garden plot. Prawer referred to these two distinct measurements as the "official" *carruca*, corresponding to the local *faddan rumi* or Western *manus*, which was the area that could be ploughed by a team of oxen in a year, and the "local" or "Saracen" *carruca*, which was the area of land that could be ploughed by a team of oxen in a single day. This is equivalent to the *faddan arabi*, which varies in size from 734m^2 in hilly regions to almost twice that in the plains (Prawer 1980:157-59).

6 As many documents refer to casalia as having both pasture and gastinae, the latter term was probably used to describe distant pasturelands or uncultivated lands (Riley-Smith 1967:434; 1973:44).

7 *Casal Imbert* (Achzib) in the north may possibly have had a similar plan (Benvenisti 1970:221-22). Only limited and unpublished excavations have been carried out at el-Bira. However, el-Qubeibeh (Bagatti 1947/1993) and more recently el-Kurum (Onn and Rapuano 1995; Boas 1996 and forthcoming), have been extensively excavated, and the recently discovered village at el-Haramiyeh (Zelinger forthcoming) has been partly excavated.

8 On Frankish tooling and masons' marks as a means of identifying Crusader period masonry, see Boas 1999:218-222.

REFERENCES

Abel, F.M.
1926 Les deux Mahomerie, el-Bireh, el-Qoubeibeh. *Revue Biblique* 35: 272-83.

Bagatti, B.
1947 *I monumenti di Emmaus il Qubeibeh e dei dintorni*. Jerusalem: Franciscan Press.
1993 *Emmaus-Qubeibeh*. English trans. by R. Bonanno. Jerusalem: Franciscan Press.

Bennett, H.S.
1960 *Life on the English Manor*. Cambridge: Cambridge Unviersity Press.

Benvenisti, M.
1970 *The Crusaders in the Holy Land*. Jerusalem: Israel Universities Press.
1982 Bovaria-Babariyya: A Frankish Residue on the Map of Palestine. Pp. 130-52 in eds. B.Z. Kedar, H.E. Mayer and R.C. Smail, *Outremer: Studies in the History of the Crusading Kingdom of Jerusalem*. Jerusalem: Yad Ben Zvi.

Boas, A.J.
1996 A Recently Discovered Frankish Village at Ramot Allon, Jerusalem. Pp. 583-94 in M. Balard (ed.), *Autour de la Première Croisade. Actes du Colloque de la Society for the Study of the Crusades and the Latin East (Clermont-Ferrand, 22-25 juin 1995)*. Paris: Publications de la Sorbonne.
1999 *Crusader Archaeology. The Material Culture of the Latin East*. London and New York: Routledge.
Forth-coming *A Frankish Village at Ramot 06, Jerusalem.*

Cahen, C.
1950-51 Notes sur l'histoire des croisades et de l'Orient latin, 2 :Le régime rural au temps de la domination franque. *Bulletin de la Faculté des lettres de Strasbourg* 29 (1950-51), 286-310.

Delaville le Roulx, ed.
1894 *Cartulaire générale de l'ordre des Hospitaliers de*

Saint-Jean de Jérusalem (1100-1310). Vol. 1. Paris.

De Vaux, R., A.M. Stéve
1950 *Fouilles à Qaryet el-`Enab, Abu-Ghosh, Palestine.* Paris: J. Gabalda.

Duby, G., ed.
1975 *Histoire de la France rurale* vol. 1. Paris: Seuil.

Ellenblum, R.
1998 *Frankish Rural Settlement in the Latin Kingdom of Jerusalem.* Cambridge: Cambridge University Press.

Ellenblum, R., Rubin, R., and Solar, G.
1996 Khirbat al-Lawza, a Frankish Farm House in the Judean Hills in Central Palestine. *Levant* 28:189-98.

Ernoul
1871 *Chronique d'Ernoul et de Bernard le Trésorier,* ed. M.L. de Mas Latrie, Paris: Libraire de la Societé de l'histoire de France.

Gil, M.
1992 *A History of Palestine 634-1099.* Cambridge: Cambridge University Press.

Kedar, B.Z.
1990 The Subjected Muslims of the Frankish Levant. Pp. 135-74 in *Muslims under Latin Rule, 1100-1300,* ed. J.M. Powell, Princeton, N.J.

Kletter, R.
1996 Jerusalem, Har Hozevim. *Excavations and Surveys in Israel* 115: 70-1.

Ibn Jubayr
1952 *Travels of Ibn Jubayr.* trans. R.J.C. Broadhurst. London: J. Cape.

May, N.
1997 Jerusalem, Har Hozevim. *Hadashot Arkheologiot* 107:81-84 (Hebrew).
2000 A Crusader Period Farmhouse at Har Hotzvim, North of Jerusalem. Hebrew, *Qadmoniot* 33/1 (119):40-45.

Onn, A. and Rapuano, Y.
1995 Jerusalem, Khirbet el-Burj. *Excavations and Surveys in Israel* 14:88-90.

Prawer, J.
1969 *Histoire du royaume Latin de Jérusalem.* 2 Vols., ParisEditions du Centre National de la Recherche Scientifique.
1972 *The Latin Kingdom of Jerusalem. European Colonialism in the Middle Ages.* London: Weidenfeld and Nicholson.
1980 *Crusader Institutions.* Oxford: Oxford University Press.

Prawer, J. and Benvenisti M.
1970 Palestine Under the Crusaders'. *Atlas of Israel.* Jerusalem: Survey of Israel.

Preston, H.G.
1903 *Rural Conditions in the Latin Kingdom of Jerusalem during the 12th and 13th Centuries.* Thesis, Philadelphia: University of Pennsylvania.

Pringle, R.D.
1983 Two Medieval Villages North of Jerusalem: Archaeological Investigations in al-Jib and ar-Ram. *Levant* 15:141-77.
1985 Magna Mahumeria (al-Bira): The Archaeology of a Frankish New Town in Palestine. Pp. 147-168 in *Crusade and Settlement: Papers Read at the First Conference of the Society for the Study of the Crusades and the Latin East and Presented to R.C. Smail* ed. P.W. Edbury. Cardiff: University College Cartidge Press.
1986 *The Red Tower (al-Burj al-Ahmar):Settlement in the Plain of Sharon at the Time of the Crusaders and Mamluks, A.D. 1099-1516. British School of Archaeology in Jerusalem, Monograph Series, Vol. I.* London.
1992 Aqua Bella: The Interpretation of a Crusader Courtyard Building. Pp. 147-67 in *The Horns of Hattin,* ed. B.Z. Kedar, Jerusalem: Yad Ben Zvi.
1993 *The Churches of the Crusading Kingdom of Jerusalem,* vol. 1. Cambridge: Cambridge University Press.
1994 Burj Bardawil and Frankish Settlement North of Ramallah in the Twelfth Century', Pp. 30-59 in *The Frankish Wars and their Influence on Palestine. Selected papers presented at Birzeit University's International Academic Conference, March 13-15, 1992.* eds. K. Athamina and R. Heacock. Bir Zeit.
1994a Towers in Crusader Palestine. *Château Gaillard: 'tudes de castellologie médiévale, Vol. XVI. Actes du colloque internationale tenu à Luxembourg.* 1992: 335-50. Caen.
1997 *Secular Buildings in the Crusader Kingdom of Jerusalem. An Archaeological Gazetteer.* Cambridge: Cambridge University Press.

Richard, J.
1985 Agricultural Conditions in the Crusader States. Pp. 251-94,in *A History of the Crusades* 5, ed. K.M. Setton, Wisconsin, Mass: University of Wisconson Press.

Riley-Smith
1967 *The Knights of St John in Jerusalem and Cyprus 1050-1310.* Cambridge: Cambridge University Press.

Runciman, S.
1952 *A History of the Crusades.* 3 Vols. Cambridge: Cambridge University Press.

Smail, R.C.
1987 *Crusading Warfare 1097-1193.* Cambridge: Cambridge University Press.

Strehlke, E.
1869 *Tabulae Ordinis Theutonici.* Reprinted 1975. Berlin.

Tafel G.L. and G.M. Thomas
1856-57 *Urkunden zur älteren Handels- und Staatsgeschichte der Republik Venedig*. Vol. II. Vienna (Reprinted, Amsterdam 1964).
William of Tyre
1986 *Chronicon*, ed. R.B.C. Huygens, in *Corpus Christianorum Continuatio Mediaevalis*. Vols. 63- 63a. Transl. E.A. Babcock and A.C. Krey. Turnholt: Brepols.
1943 *A History of Deeds Done Beyond the Sea*. 2 Vols. New York: Columbia University.
Zelinger, Y.
Forthcoming *A Crusader Village in Wadi el-Haramiyyah.*

Figure 9-1: Plan of street villages in the Kingdom of Jerusalem. 1) el-Kurum; 2) *Parva Mahumeria* (el-Qubeibeh).

Figure 9-2: View of the street village of el-Kurum northwest of Jerusalem.

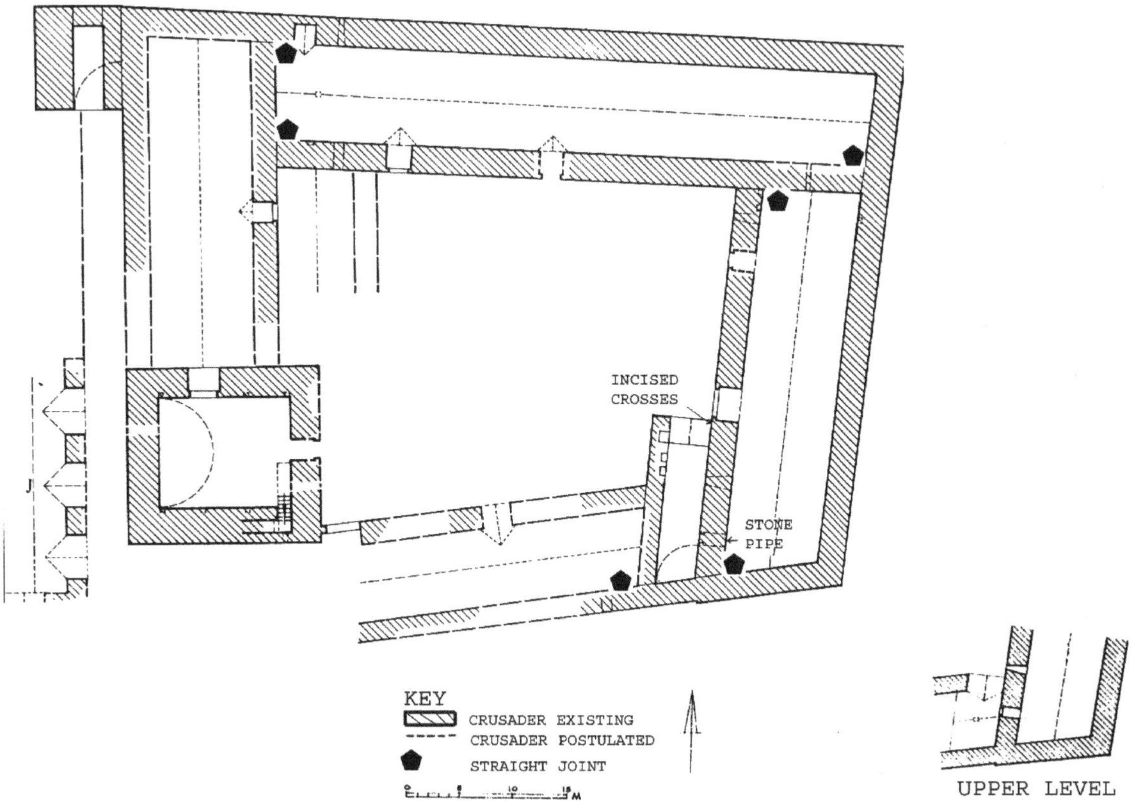

Figure 9-3: Plan of the Frankish manor house at ar-Ram (after Pringle 1983: fig. 6).

Figure 9-4: View of the Frankish manor house at Aqua Bella (view from the south-east).

Transformations in the Agriculture of al-Sham during the Mamluk period (1250-1517 CE)

ZOHAR AMAR

The Martin (Szusz) Department
of the Land of Israel Studies
and Archaeology
Bar Ilan University
Ramat Gan, 52900
Israel
amarzo@mail.biu.ac.il

A process of decline can be seen in the agriculture in al-Sham in the later Middle Ages. This process was caused by a variety of reasons, the major being the listless, uninspired policy of the Mamluk rulers, coupled with the severe damage that they inflicted on the peasant class. The decline was a gradual process, whose effect was fully felt only late in the Mamluk period. In the following study, this process will be discussed, based on an examination of the evidence for the cultivation of various agricultural crops during this period, such as grapes, date palms, grains, and other fruits and specialized crops.
Concurrent with this decline in al-Sham, at the end of the fourteenth century CE there was a commercial and industrial awakening in Europe, and the technological and commercial initiative shifted to the West. Due to these changes, markets in the East were flooded with agricultural products from the West, further compounding the reduced production and commercial abilities in al-Sham during the Mamluk period.

Agriculture in al-Sham (Israel and Syria) during the late Middle Ages (Amar 1992: 229-334; Fraenkel 1995) is marked by a dramatic turn for the worse. As early as the Ayyubid period the first signs of erosion in the status of the peasant population in particular and a crisis in agriculture in general had already appeared. In the Mamluk period, however, these trends continued, intensified, and took on their final shape. This was a gradual process that left a marked impression by the end of the Mamluk period which will be illustrated in this article through a study of agricultural crops. In this context, however, it is important to characterize the policy of the Mamluk state towards agriculture and the economy.

The Policy of the Rulers Towards the Middle Class

At the beginning of the Ayyubid period (1187 CE) the Land of Israel was in effect divided up into two parts: the western sector, including the Lebanese coastal plain until the region of Jaffa, was under Crusader-Christian control and the remainder was under Ayyubid-Moslem rule. After the death of Saladin (1193 CE), the Ayyubid dynasty underwent a process of disintegration as a result of the disputes and power struggles among his heirs. Al-Sham became an arena of continuous battles, and during this period, agreements and pacts were signed both between the Crusaders and the Moslems as well as among the various Ayyubid rulers who were themselves divided. Over this short period, lands and villages frequently passed from hand to hand, and consequently, there was no continuity in the policy guiding their political subordination.

The most significant change in the attitude towards agriculture stemmed from the lack of interest and concern on the part of the Ayyubid and Mamluk rulers regarding the state of affairs in the Land of Israel. In contrast to the Crusaders, who had a direct, positive involvement in the economic life of the country, the later rulers controlled the area from their distant centers of Cairo or Damascus, while this applied through almost all of the First Muslim Period (638-1099 CE). The beginning of the Ayyubid period represented was a decisive turning point regarding the status of the peasantry specifically and agriculture in general. Until the Ayyubid period, the central regime controlled the rural regions by means of an apparatus that had already been devised in the Byzantine period. In this ancient system, the governing apparatus was directly involved in agricultural life, controlling crops, techniques, and processing methods, maintenance, water rights, and the like. Under the traditional method of supervision, rulers had the opportunity (if they so desired) to rebuild and develop large agricultural endeavors. From the Ayyubid period on, however, there was a gradual decline in the quality of supervision, services, and degree of enforcement that had existed under the aegis of the ancient governing apparatus. This is linked to the material change in attitude towards the agrarian class in the Land of Israel. This broad and complex issue is beyond the scope of this paper (see, however, Poliak 1939), rather, an analysis which will focus on one of the important factors: the method of tax revenue from collection known as "Iqta' ". This method is mentioned in several Moslem sources from earlier periods (EI^2 s.v. Ikta; Cahen 1953:25-52; al-Duri 1969:3-22), and became a dominant factor with a negative impact in this region only in the late Middle Ages. According to Cahen (1953: 45-52) the change for the worse had already begun in the Ayyubid period and according to Irwin (1977: 64,72), its effect is only noticeable from the beginning of the fourteenth century. In this author's opinion, the decisive critical point is rooted in the fact of the final termination of Crusader hegemony in the Land of Israel in 1291 CE.

According to al-Maqrizi (1364-1442 CE), the Iqta' system was used only sporadically as a method of collecting revenue for paying the troops and for other state expenditures, but since the time of Saladin, the Iqta' became official practice, employed systematically and regularly. Under the Iqta' system in its later format, as adopted by Saladin in Syria and Egypt (al-Maqrizi 1911:96-7), the land belongs to the state, which grants army officers the right to collect taxes on a given piece of land. With the expansion of the institution of the Iqta', central control effectively came to an end. This system destroyed the last remnants of the old governing apparatus, creating in its place two major systems of collection: the Iqta' system and that of the regime itself. Under the Iqta' system, the concession holders lived far away from the lands to which they held rights and used a remote control method of management (Poliak 1939:18-31; Poliak 1940:36-9). It was a predatory apparatus of collection lacking any system of control, which treated the peasantry with unbridled brutality and high-handedness. This also stemmed from the desire of the concession holders to cover the procurement expenses of the office and job holders under their authority within a short period of time, since the concessions were, in fact, provisional. Concession holders took advantage of their military status and often collected more than their due. Conflicts of interest frequently arose between the government and the military commanders, who unlawfully helped themselves to a share of the tax that rightfully belonged to the government. This led to an immediate drop in state revenue, and as the situation became more difficult, the state was forced to expand the Iqta' system to compensate for its losses.

The situation of the peasantry under the government collection system was no better. This period witnessed a process of supplanting the small land-holder class through increasing the involvement of the regime not only in those land-holdings that belonged to it directly (on the Sultan's land-holdings, see Poliak 1940:28-30), but also in all the areas under its jurisdiction. Subsequently, this meant dispossessing the peasants from their lands and gradually turning them into serfs. Legally, the status of the peasant was inferior even to that of a slave (Poliak 1940:57; al-Maqrizi 1911:85-6). In addition to the blow to the status of the peasant, the Mamluk regime exercised over-involvement in economic life and trade in the Land of Israel, suppressing almost any possibility of conducting free economic activity. Many products were considered a government monopoly, and almost all basic crops were under the direct control of the authorities. The regime fixed the prices of products, sometimes at an inflated rate ('*tarh*') and the public was forced to buy the products at these prices. The apathy of the Mamluk regime to the situation in the country is reflected, among others, in the its failure to initiate any large agricultural enterprises and in ignoring the needs of the peasants. The regime also showed no foresight in developing and introducing new technologies and techniques. The Mamluks' conservatism was ultimately their downfall. Their exploitation of the peasantry and the lack of incentives and initiatives from the representatives of the regime quelled all positive potential and motivation on the part of the peasants, who moved over to a basic subsistence economy. The stagnation of the economic system led to the atrophy and damage of the economic limb on which they were sitting. Later, their lack of openness to innovations, such as the introduction of modern firearms into the army, led to the defeat of the Mamluks on the battlefield by the Ottomans and their disappearance from the political map.

Economic Policy

The policy of the Mamluks should be examined both from the agricultural perspective and in the context of their overall economic theory. The Mamluk regime conducted a policy characterized by passive, unimaginative administration, side by side with the exploitation of the resources of the country as a source of revenue for the state coffers. This short-sighted economic view created irreversible damage in the long run for the agriculture of the Land of Israel and ultimately impaired the revenue of the regime itself. J. Drory provides a succinct description of the characteristics of Mamluk economy policy:

> "The Mamluks were blinded by the wealth of the country and administered an unbridled and uncontrolled economy which was wasteful and profligate, investing in public works with no guarantee of their financial stability; showing a passive and unenterprising attitude to the commercial and productive forces in the Mamluk state. They were satisfied to exploit the wealth readily-accessible to them by virtue of their monopolistic geographic status but made no effort to seek out new markets or methods to make trade more efficient. The parasitic approach taken by those who charted economic policy led to its collapse in the face of competition with economic

forces which were more dynamic and energetic." (Drory 1979:119-20).

In addition, the institutions of the Waqf gradually grew stronger and expanded in the Land of Israel in general and in Jerusalem in particular, beginning during the time of Saladin through the Mamluk and Ottoman periods. The status of the Waqf officials who seized administrative and governing positions became stronger (Fraenkel 1992; Shoshan 1992:86-97). As the result of the transformation of Jerusalem into a religious center and neglecting it administratively, the management of its economy came both directly and consequentially under the control of the religious officials. Expanding the religious system and staffing it with office holders who received salaries or various emoluments and annuities led to the creation of a non-productive economic administration which was open to corruption and moral pollution. Moreover, since the property of the Waqf was 'sacred', it was immune to governmental encroachment and exempt from taxation. The Waqf and its growth greatly contributed to the neglect and negligence in managing the land and thus constituted an additional factor that further diminished the state's revenue.

This portrayal of the characteristics of Mamluk policy is, of course, very general and does not represent the specific reality that prevailed in each generation, for there also were individual rulers who contributed to the development of the al-Sham. Generally, however, one may point to two crucial points in the decline of agriculture in the al-Sham. One is connected with the severe economic crisis suffered by the entire Mamluk empire at the beginning of the fifteenth century with the rise of the Mamluk rulers of Cherkassian origin (Ashtor 1964:160, 164). This crisis was accompanied also by internecine ethnic rivalry, which weakened the power of the central authority. The second turning point came at the end of the Mamluk period, with the undermining of the general stability in the country in the context of the Bedouins' increasing power. The extent of influence of the Mamluk regime on the agriculture of the Land of Israel also differed in various parts of the country; the coastal area throughout the period was especially damaged by the destructive policy of the Mamluks, as a result of their maritime vulnerability and their fear of a European invasion (Ayalon 1983:147-52).

While some of the country's interior regions maintained a reasonable degree of economic stability, it seems, however, that even the accepted theory that the destruction of the coastal cities in 1291 CE caused damage to their agricultural and economic backbone warrants re-examination. It appears that the force of this process was not as decisive as previously considered. The evidence of lively commerce and trade conducted in the coastal cities of the Land of Israel (among these Acre and Jaffa) during the fifteenth century demonstrates the activity of these port cities already at the end of the Mamluk period, earlier than scholars had previously suggested. In Acre part of the old city was re-populated as early as the fifteenth century (Arbel 1988:238-239, 256-257). Around the ruins of Jaffa, a large rural nucleus sprang up, which formed the basis of the city of Jaffa as it was known in the Ottoman period. In the environs of Jaffa, the peasants grew grain, cotton, and watermelons which became especially renowned in later periods (Suriano 1949:38). The town of Magdal took the place of Ashkelon which had been destroyed and as early as the beginning of the sixteenth century was again mentioned as an important cotton-growing center (Hamarneh 1978:253). A considerable part in this process of renewal must be attributed to the extensive involvement of Italian merchants in the country's coastal cities and the increase in the pilgrim traffic. While this traffic, which was also touristically-oriented, made only a peripheral contribution to the development of the country, it brought in a great deal of money and represented a source of livelihood for many people. Due to the involvement of the regime into whose coffers most of the money from the pilgrim industry was paid, the pilgrims' visits did not contribute greatly to the welfare of the population of the Land of Israel.

Political and Historical Factors

Ultimately the downward spirals referred to above were the result of a complex series of processes that took place in the Mamluk empire. The decline in agriculture in the Land of Israel may also be attributed to other political and historical factors that were not under the direct control of the Mamluk regime, such as the impact of the 'Black Death' in the middle of the fourteenth century (al-Maqrizi 1973:772-75; Dols 1977), the sharp demographic decline that affected all of the countries of the Near East (Ashtor 1981a:253-57), the Mongolian invasions at the end of the thirteenth and beginning of the fourteenth centuries, the last of these under the leadership of Timur Lenk (1400 CE), and the rise of the Ottoman Empire. Although I agree with the premise that these factors had a substantial, significant, and cumulative effect on the conduct of normal life in the Land of Israel, it is difficult to attribute to them alone the decisive and specific cause of the setback suffered by agriculture. In my opinion, the decisive factor in this case was the policy implemented in the Land of Israel. Under the assumption that under these kind of normal agrarian conditions, cultivated by an efficient regime, the country could have extricated itself from these kinds of crises over time (Amar 1997a). Thus, for example, the ability to halt the Mongolian threat at the beginning of the Mamluk period (1260) attests, to a large extent, to the staying power of the emerging Mamluk kingdom. Even after the invasion by Timur Lenk, however, there is evidence showing the vast scope of commerce that was conducted between Damascus and Europe by the Italian merchants. In other words, the volume of trade did not diminish, and the blame for the decline in agriculture and industry (such as sugar and paper) must be laid squarely at the door of the policies conducted by the Mamluk regime (Ashtor

1977:273). Moreover, some scholars see the discovery of the Cape of Good Hope route at the end of the fifteenth century as an additional decisive factor that materially damaged the economy of the Mamluk state (e.g., Heyd 1936: 427). This view is not unequivocal, however, for a lively, regular spice trade continued between the Far East and Europe along the traditional route, through Egypt and the Mediterranean (Inalcik 1970:331-32). At the same time, it is clear that the discovery of the Cape of Good Hope did away with the monopolistic status that the Mamluks had enjoyed until then.

The Collapse of Agriculture at the End of the Mamluk Period

The characteristics of Mamluk policy. the first signs of which were already beginning to emerge in the Ayyubid Period, led to a profound crisis in the agrarian culture of the Land of Israel at the end of the Mamluk period. This crisis is the almost inevitable result of the factors enumerated above and which affected, either directly or consequentially, the various agricultural industries. These were prolonged and extended processes which varied from region to region and even from crop to crop. The limited sources available to us do not allow us to trace these processes with any accuracy and determine their degree and scope. In most cases, the changes can be discerned only in a general way and beginnings and ends established. In any event, **an analysis of the sources describing the agricultural crops grown in the Land of Israel in the Middle Ages unequivocally shows that at the end of the Mamluk period, the agricultural infrastructure of the country and most branches of agriculture were damaged.** A survey of the condition of the principal basic and other crops is presented below.

The Date and Sugar Industries

Historical and archaeological sources provide clear evidence of the existence of a large sugar production industry in the area of the Jordan Valley, which was the result of official government initiative and direction. This large-scale industry apparently did not exist in the region before the Ayyubid period and was abandoned at the end of the Mamluk period (Franken and Kalsbeek 1975: 143-54; Ibrahim, Sauer and Yassine 1976; Hamarneh 1978:12-19; Amar 1997b:313-16). Alongside the rise of the sugar industry there is a discernible downward trend in the date industry, until it disappears completely from the region and from the landscape of the Land of Israel. This is in contrast to the impression that emerges from the sources, according to which dates were a major crop in all the thermophilic regions of the country, at least until the Crusader period. These two processes cannot be ignored, despite the fact that the sources do not explicitly state the connection between them. Dates are mentioned in the sources as an important agricultural crop in the Jordan Valley at least until the thirteenth century, while sugar became a major industry only from this century on. It was a gradual process first discerned in the Ayyubid period in the Beth Shean region. The city had been famous for its dates at least until the twelfth century (see, e.g., al-Muqaddasi 1906:162,180; al-Idrisi 1974:356), but already in the Ayyubid period, Yaqut attests that he personally had never seen more than two date palms bearing fruit (1866-70: I, 788). On the other hand, he states that Beth Shean was the most important city in the Jordan Valley after Tiberias, and primarily sugar is grown there (Yaqut, 1866-70: IV, 51). The city served as a center for the regime's business transactions throughout the Jordan Valley. In this period all the sugar producing centers in the region were under the direct control of or owned by the regime (al-Qalqashandi 1915: IV, 183, 188). These sources only note the beginning of the process by which the date industry suffered in contrast to the development of the sugar industry, a process that accelerated and reached its peak in the Mamluk period.

The disappearance of the date palm from the general landscape of the Land of Israel was a gradual process. In ca. 1322 CE, Estori ha-Parchi wrote: “We also have here in the Land of Canaan two kinds of drinks, which due to their thickness cannot mix with each other, except after a long period of time, like date honey” (Estori ha-Parchi 1897-99:454). This reality had already changed by the end of the fourteenth century when R. Yosef Tov-‘Alem (1370 CE), who lived in the Land of Israel for many years, attests that “there is no date honey there at all nowadays” (Tov-‘Alem 1930:9). Al-Qalqashandi (d. 1418 CE) also writes: “The agricultural produce in al-Sham is unlike that of Egypt and one cannot find there the ‘balah’ [a species of dried date, Z.A.] or ‘ratab’ [a species of moist date, Z.A.]” (1915: IV, 87). Indeed, no mention is made in one Italian notary's documents of commerce in dates in the Land of Israel and most of the date exports from the east to Europe came from Egypt (Ashtor 1987:35, 53). In the Mamluk period it appears that dates were even imported into the country. Testimony of the import of dates to Ramla is recorded in the Ottoman archives of the sixteenth century (Cohen and Lewis 1978:55, 141).

There is also documentation of the gradual damage to the date industry with reference to the northern coastal region. At the beginning of the Ottoman period (1535 CE) R. David de Rossi testifies that in Tripoli: “Dates are cheap but the dates are not good like those that come from Egypt” (Ya'ari 1943:186). Rauwolf, too, noted in 1573 CE, that in Tripoli, there were a few lowly date trees (1971:24). Dates had disappeared completely from the Jordan Valley at the end of the Mamluk period, as noted by R. Ovadiya of Bartenora, incidental to an account of the economic situation in Jerusalem:

"However, I have not found date honey here, nor have I seen any dates themselves, and in Jericho, the City of Dates, a reliable person who was there told me ... that in all of Jericho there are only three poorly date palm trees and these do not bear fruit" (Ya'ari 1943:132).

There is also evidence of date palm orchards dwindling during the Mamluk period in the region along the Suez Canal, identified by Crusaders as the Biblical "Eilam". Ya'akov of Verona wrote in 1335 CE that there were more than ten thousand date trees in this location (Rohricht 1895:237). At the end of the Mamluk period, however, another pilgrim estimated the number of date trees at approximately two thousand (von Groote 1860:132).

The explanation proposed here sees the expansion of sugar growing and the disappearance of date cultivation as proof positive of the process of decline in the agriculture of the Land of Israel, particularly in terms of irrigated fields, as a result of the official policy of the regime. It would appear that intensive and deliberate development by the Ayyubid and Mamluk regimes of agriculture in the Jordan Valley should be considered a positive development that contributed greatly to the flourishing of agriculture in the region, but there is another side to the coin. The agricultural development of the region was limited to only sugar growing and did not apply to any other crop. The motivation was not a concern to develop the region but rather a desire to be rewarded with large, easy profits to line individual pockets, for these lands belonged to the sultan and his emirs, not to the state treasury. Sugar cane was the most appropriate crop for this purpose, since it grew quickly and its cultivation did not require complex agricultural technology, in comparison with the date industry, which required know-how and trained manpower over its prolonged growing period. Incidentally, it must be noted that new crops are mentioned in the sources : citrus fruits and eggplants appear for the first time in the Jordan valley only during this period. Although these crops did not have very high economic value, it seems that the extension of their cultivation into the region was made possible by the agricultural development of that area at the time (Amar 1997b:325-26). According to Ashtor, the short-sighted policy of the authorities (which sought immediate gain and shunned long-term investment) eventually also led to the collapse of the flourishing sugar industry. The cause of the decline in the sugar (and paper) industries in the Levant was stagnation in local technology, as opposed to the process of development and technological innovation in the West (Ashtor 1981b:105-7). The Venetian sugar trade in particular and the European sugar trade in general which, until the beginning of the fifteenth century, were based primarily on the industry in the Levant, began to look towards the trade centers of Europe. Furthermore, at the beginning of the sixteenth century, the Land of Israel was no longer able to supply its own needs for sugar and was forced to import it.

Olives

Olives are one of the basic crops that was barely affected during the Mamluk period, because the trees are hardy and do not require a great deal of constant cultivation, and because growing olives is deep-rooted within the local agrarian culture. From the first half of the Mamluk period, there is even evidence of the export of olive oil and olive-oil soap from al-Sham, primarily from Nablus to various countries, among them Egypt, Hejaz, and Yemen (al-Dimashqi 1866:200; al-Qalqashandi 1915: IV, 87; al-Umari 1985:26). Apparently it was only in the central and southern regions that some of the olive groves thinned out with the destruction of the coastal plain cities during the Mamluk period. In contrast, the olive industry became predominant in the landscape of Samaria only during this period, in parallel to the decline of the grape industry (Amar 1995).

The olive industry, however, did not entirely escape the corruption of the Mamluk policy. Towards the end of the regime, the number of testimonies of heavy levies imposed on olive oil increases. In many places in the Land of Israel the soap manufacturers and exporters were forced to purchase olive oil at the price and of the quality determined by the authorities, which treated locals harshly. The method of transactions carried out under duress and at inflated prices (*tarh*) was current in many places in al-Sham: Nablus (Mujir al-Din 1973:356-58; 365-66; 373-74), Tripoli (Van-Berchem 1909:59, 77), and Jerusalem (Ya'ari 1943:158). This harsh policy hurt the peasants, and many olive groves were abandoned. Thus, Felix Fabri for example, notes in his account of his journey to the Hebron Valley that before his entourage even entered the city of Hebron he saw stone-terraced hills, now neglected, which in the past had been the site of the cultivation of, among other crops, olive trees. These were now replaced by thorny weeds (PPTS 9-19:408). Evidence of damage to the olive industry and its products appears in several documents on the import of olive oil and soap from Europe into Syria by Italian merchants in the fifteenth century. It transpires that the cost of olive oil originating in the south of Europe or Tunisia was low, at times even one hundred percent lower than the price of local olive oil. There is, for example, documentation of a Venetian ship that brought olive oil worth 25,000 ducats into Syria in 1406 CE (Ashtor 1981a:258-59). Soap was also imported into the region, including Syria from the Italian cities and from Cyprus (Ashtor 1977: 73, No. 2; Ashtor 1981a:277-78). It is possible that both the quality and value of local soap declined, and the consumers were forced to import cheaper soap of better quality of European manufacture, which used better-developed technology.

Wine Grapes

In comparison with the olive industry, which managed to preserve its status almost throughout the Mamluk period, there was a turn for the worse in the grape industry which until this period had been one of the most important agricultural products of local, traditional culture. The decisive change is characterized by a reduction in the number of vineyards that produced grapes for the wine industry rather than for table grapes used in the production of raisins and '*dibs*' (grape honey) which only at this time became a common and basic food product, a substitute for the wine prohibited by Islamic law (Amar 1995; Amar 1996:241-48). This transformation was a direct result of the Islamization of the Land of Israel in the Mamluk period. This new trend was described by Burchard of Mount Sion ca. 1283 CE:

"There are many grapevines in the Holy Land and there could have been more but the Saracens who rule the land now do not drink wine, except a few among them - in stealth- and they are destroying the vineyards, except for those living close by the Christians who raise them for profit, so that they may sell them to the Christians..." (Laurent 1864:88).

Indeed only from the Mamluk period are there numerous testimonies of religious zealotry among the Moslem population against their wine-drinking protected subjects, a phenomenon that previously had practically been non-existent in the Land of Israel. The Mamluk rulers, who exhibited a marked religious devoutness, took strong measures, not always with complete success, to put an end to the commerce in and use of wine among Moslems. The Mamluk regime prohibited the drinking of wine in public and even restricted the production of wine by its protected non-Moslem subjects. The Christians and Jews continued to cultivate wine grapes to a limited extent, after receiving personal concessions and under the strict supervision of the authorities. The harsh restrictions imposed by the authorities led to an increase in the price of wine, as attested by many pilgrims (e.g., Laurent 1864:88; Newett 1907:238; Letts 1953:51). Granting a license to manufacture the forbidden beverage without restriction, in exchange for payment to the authorities, brought abundant revenue to the state coffer and to local officials who turned it to their personal profit . These authorities cruelly exploited and plundered their protected subjects, who were like hostages, since wine was an essential basic product, vital to their subsistence (Ya'ari 1943:95, 136, 156).

The greatest regional change affecting the distribution of vineyards in the Land of Israel occurred in Samaria during the Mamluk period. The grape industry, which together with the olive industry, had predominated in the agricultural landscape until the Crusader period, almost completely disappeared during the Mamluk period, and in its place, the olive industry grew and became the central, nearly exclusive, agricultural crop in Samaria. Even during this period, however, the olive industry did not entirely replace grape-growing in most of the Mediterranean hilly regions, like the Galilee and Judea. It may be assumed that with the increasing Islamization, more land was allocated to olive cultivation at the expense of grape-growing, but the impression is that grapes and olives remained the two primary agricultural crops, similar to the situation in earlier periods, with the olive industry enjoying a certain advantage (Amar 1995).

Grain

Regarding the cultivation of wheat, the impression emerges that during the Mamluk period, al-Sham in general continued to meet its own demand, although at the same time, there is a reduction in grain cultivation in comparison with earlier periods. It appears that in the Mamluk period the country shifted to a closed economy. Land allocated for growing grain was restricted, only supplying the subsistence needs of the local population. In the beginning of the Mamluk period al-Sham still represented the "food basket" of Egypt. Thus, for example, large quantities of wheat were transferred to Egypt during the drought years of 1295 CE and 1336 CE from Karak, Shobak, Gaza, and the region around Damascus (al-Maqrizi 1973:394; Allouche 1994:44). However, according to the research carried out by Ashtor, a serious economic crisis broke out in the countries of the Mamluk empire at the beginning of the fifteenth century and brought in its wake a steep rise in prices. The price of wheat rose by 20% and the price of bread by 30%. The crisis became more pronounced with the rise of the Cherkassian Mamluks (Ashtor 1964:160-64). During the years of drought and famine that plagued the Near East, there was extensive importing of wheat from southern Europe to Syria and Egypt through the offices of the Italian merchants (Ashtor 1984).

Suriano wrote at the end of the Mamluk period that the local residents were not making full use of the potential of the land but were satisfied with minimal cultivation to meet their own needs (Suriano 1949:09), known today as subsistence farming. Ashtor is of the opinion that due to the dwindling of the population of the Near East, the price of grain dropped on the one hand, while on the other, cotton growing became more widespread in the Land of Israel (Ashtor 1986:109).

Wheat, as well as dates and wine grapes gave way to other crops such as olive trees and sugar cane which were more profitable. Although these industries required considerable manpower, the crops were better able to withstand adverse conditions of nature. An additional explanation for the change in principal crops is the predatory policy of the regime towards its subjects. As in most branches of agriculture, the Mamluk regime was involved in fixing the prices of wheat and barley (see, e.g., Ibn Tulun 1964:91), and most of the revenues from agriculture went into the sultan's treasury and into the pockets of its corrupt officials. Insofar as it depended on the peasants' preferences, it was not worthwhile for them to grow wheat beyond what they required for their own needs, and then they often preferred to grow barley instead.

Cotton and Mulberry

Cotton appears to be one of the few agricultural crops that continued to develop in the Land of Israel throughout the Middle Ages. It caught on following the Islamic conquests and took on considerable importance in the textile industry and trade, in both the East and West (Watson 1983:31-41; Amar 1998:40-3). As opposed to other basic crops, such as sugar, the cotton industry was not hurt in any significant way during the Mamluk period. On the contrary, it appears that it was during this of all periods cotton took its place in the market as the leading export product.

The demand in Europe for cotton from the East undoubtedly led to the expansion of the cotton growing areas in the Land of Israel during the later Middle Ages. After the fall of Acre and other Crusader cities, when the peasants in these areas lost an important market for their produce, many shifted to cotton growing. Moreover, when grain prices in the Near East dropped, due to the reduction in population, the quantity of cotton raised in the Land of Israel apparently increased again. In other words, with the demand for grain considerably reduced, the agricultural sector found an alternative cash crop that was well suited to the demands of the financial market (Ashtor 1976:712; Ashtor 1986:108-9). It would appear that the explanation for this phenomenon is that the Mamluk regime was not directly involved in cotton growing and the cotton trade in the same crucial way it had been involved in the agricultural production of sugar, for example. While the regime did impose taxes on the peasants and various levies on the Venetian merchants, cotton ultimately remained a profitable crop. The monopoly on the cotton trade was in effect held by the European merchants, particularly by the Italians. This situation was the result of a complex relationship built over a long period of time between the Mamluk regime and these merchants (Wansbrough 1965), which may have even begun with their Ayyubid predecessors. The expansion of the cotton industry was accompanied by the gradual development of a well-organized marketing infrastructure. The Venetian merchants were engaged in the trade in cotton and its products throughout al-Sham, including peripheral areas, with a distribution of agents at key points like Tripoli, Acre, and Ramla, and thus effectively created a network of trade throughout the Land of Israel (Ashtor 1976:690-98). The Italian merchants purchased the cotton directly from the local peasants, and the prices were fixed within the framework of a free market based on the law of supply and demand (see, for example, a document from 1497 in Wansbrough 1965:509-23; Arbel 1988:246, 258). If this were the situation in the other branches of industry and agriculture, the economy of the Land of Israel at the end of the Mamluk period would probably have been quite different. The intensive commerce between the European coastal cities and the Levant continued unabated during this period, thus proving that the cultural and political rivalries between Islam and the Christian world did not harm their commercial relations.

The cultivation of the white mulberry also continued to develop and grow during the Mamluk period, particularly in the north of the country. As in the case of cotton, there is no evidence pointing to significant damage to white mulberry cultivation in this period. The explanation appears to lie with the fact that this crop was closely linked to the silkworm industry, in which the Italians, who conducted an enlightened economic policy, were intensely involved (Amar 1998:48-51).

Other Crops

In addition to the basic crops discussed above, the agricultural crisis that beset the country at the end of the Mamluk period adversely affected many other crops. The first to be affected were the less common and relatively prestigious agricultural crops such as cherries, roses (for rose-water), and chestnuts. Crops that had previously been commonplace in the landscape of the country also became scarce: bananas became an expensive luxury fruit (Suriano 1949:223) and the groves of terebinth trees in the Hebron Hills stood untended. Fabri reports that the terraced fields that in the past had been planted with grapevines, olive tree, oranges, pomegranates, and other fruit trees, in his time were covered with thorny bushes (PPTS 9-10:408). The lack of many types of fruits, such as nuts, apples, pears, and almonds, was felt in the marketplaces of the Land of Israel, and merchants were forced to import them from the area of Damascus and sell them at a very high price. This also applied to local produce such as indigo and figs, that in the past had been famous for its fine quality, but now was of such poor quality that the residents imported superior varieties from neighboring countries (see, for example, Ya'ari 1943:132,172; Suriano 1949:39). Rice and henna continued to be cultivated in al-Sham, but at the same time, there is testimony that these were also imported from abroad, primarily from Egypt. Lentils, peas, and safflower were similarly imported from Egypt (Ibn Iyas 1960-1985: I/1, 42-43) which does not mean that these and other crops disappeared entirely, but it is quite clear that the profound economic crisis brought about a significant reduction in local, quality produce which did no more than satisfy local demand. In other words, the fact that previously typical and high quality local products began to be imported is evidence of the agricultural crisis.

In conjunction with the import of agricultural products from Egypt at the end of the Mamluk period, there is evidence of difficulties in the transfer of export products from al-Sham to Egypt. The export of many products such as rose water which was regularly sent to Egypt was terminated, for example, in 1501 CE, due to the uprisings and poor security arrangements in the region (Ibn Iyas 1960-1985: III, 494). The information on the transfer of seedlings of various species of Syrian fruits and flowers to Egypt in 1507 CE (Ibn Iyas 1960-1985 IV:102) is exceptional and does not reflect the typical reality in the Land of Israel. It can be surmised that these plants originated in the Damascus region, and moreover, that this was an unusual, private initiative of the Mamluk sultan, who imported these plants for his own private garden.

Summary

Towards the end of the fourteenth century, an industrial awakening and flourishing occurred in Europe, while at the same time, deterioration and enervation were plaguing the Mamluk Empire. Technological development in the West released the European consumer from dependence on imports in almost every branch of commerce - except for cotton as a raw material - that had previously been dominated by the eastern countries. Technological and commercial enterprise shifted to the West and by means of the Italian trade which dominated the Mediterranean region unrestrictedly, the markets of the Near East were flooded with cheap, quality products. In some cases this involved the export products of vegetable origin, both basic goods and luxury items, for which al-Sham had been renowned in the First Muslim Period, such as olive oil and soap, paper, and raisins (Ashtor 1981a:253-86). As we have seen, the decline in agriculture in the Land of Israel and Syria may be attributed, first and foremost, to the deleterious economic policy of the Mamluk rulers.

Some of the changes that took place in the crops of the Land of Israel in the Mamluk period may be explained in the context of changes in the gastronomic tastes of consumers. The impression is that the consumption of rice and sesame seed oil had a larger share of the basic foodstuffs among the local population than it had had in earlier times. Instead of date honey, the populace preferred to use other types of sweeteners, such as sugar and carobs. This may be a mistaken impression that stems from the wealth of sources in this period as opposed to the relative paucity of sources from earlier periods. It appears, nevertheless, that the changes in the consumption habits were the direct result of time-dependent changes. In other words, they resulted from the scarcity of certain products and the search for alternatives or, in the case of sesame, for example, a certain crop was better suited to the circumstances at a time when agriculture was in decline.

REFERENCES

Allouche, A.
1994 *Mamluk Economics. A Study and Translation of Al-Maqrizi's Ighathah.* Salt Lake City: University of Utah Press.

Amar, Z.
1992 Notes on Flora and Agriculture in Mamluk Palestine. Pp. 220-36 in *Palestine in the Mamluk Period.* ed. J.Drory. Jerusalem: Yad Izhak Ben-Zvi. (Hebrew).
1995 Grape Cultivation and Wine Production in Judea and Samaria during the Middle-Ages. Pp. 247-61 in *Judea and Samaria Research Studies* IV, eds. Z. H. Erlich and Y. Eshel. Kedumim-Ariel: Eretz Press (Hebrew).
1996 'Dibs' in the Land of Israel during the Middle Ages. Pp. 241-48 in *Judea and Samaria Research Studies* V, ed. Y. Eshel. Kedumim-Ariel: Eretz Press. (Hebrew).
1997a Man's Attitude Towards the Vegetation and Agriculture of the Land of Israel during the Middle Ages. Pp. 167-78 in *Hikrei Eretz - Studies in the History of the Land of Israel Dedicated to Prof. Yehuda Feliks*, eds. Y. Friedman, Z. Safrai and J. Schwartz. Ramat-Gan: Bar-Ilan University (Hebrew).
1997b Agricultural Products in the Jordan Valley during the Middle Ages. Pp. 297-326 in *The Village in Ancient Israel*, eds. S. Dar and Z. Safrai. Tel Aviv: Eretz Press (Hebrew).
1998 The Revolution in Textiles in Eretz Israel and Syria in the Middle Ages. *Cathedra* 87:37-60 (Hebrew).

Arbel, B.
1988 Venetian Trade in Fifteenth-Century Acre: The Letters of Francesco Bevilaqua (1471-1472). *Asian and African Studies* 22:227- 88.

Ashtor, E.
1964 Le cout de la via en Palestine au Moyen Ages. *Eretz-Israel* 7:154-64.
1976 The Venetian Cotton Trade in Syria in the Later Middle Ages. *Studi Medievali* 17:675-715.
1977 Levantine Sugar Industry in the Later Middle Ages - An Example of Technological Decline. *Israel Oriental Studies* 7:226-80.
1981a The Economic Decline of the Middle East during the Later Middle Ages: an Outline. *Asian and African Studies* 15:253-86.
1981b Levantine Sugar Industry in the Late Middle Ages. Pp. 91-132 in *The Islamic Middle East, 700-1900: Studies in Economic and Social History,* ed. A. L. Udovitch. New Jersey: Darwin Press.
1984 The Wheat Supply of the Mamluk Kingdom. *Asian and African Studies* 18:283-95.
1986 *East-West Trade in the Medieval Mediterranean*, ed. B. Z. Kedar. London: Publish Variorum Reprints.
1987 The Crusader Kingdom and the Levant Trade. Pp. 30-54 in *The Crusaders in their Kingdom,*

ed. B.Z. Kedar. Jerusalem: Yad Izhak Ben-Zvi. (Hebrew).

Ayalon, D.
1983 The Mamluks and Maritime Power as a chapter in the struggle between Islam and Europe. Pp.147-52 in *The Geography of the Mediterranean Basin*, eds. Y. Karmon, A. Shmueli and G. Horowitz. Tel Aviv: Eretz Press (Hebrew).

Cahen, C.
1953 L'Evolution de l'Iqta' du IXe au XIIIe Siecle. *Annales Economies Societes Civilisations* 8:25-52.

Cohen, A. and Lewis, B.
1978 *Population and Revenue in the Towns of Palestine in the Sixteenth Century.* Princeton: Princeton University.

al-Dimashqi, M.
1866 *Kitab Nukhbat al-Dahr fi 'Aja'ib al-Bar wa'l-Bahr,* ed, M.A.F. Mehren. Saint Petersbourg: Imprimerie de l'Academie Imperiale des Sciences.

Dols, M.W.
1977 *The Black Death in the Middle East.* Princeton, NJ: Princeton University.

Drory, J.
1979 The Attitude of Islam to the Plague in the Middle Ages. Pp. 118-21 in *Hamizrach Ha'hadash* 28:118 - 121 (Hebrew).

al-Duri, Abdul Aziz.
1969 The Origins of Iqta' in Islam, *Al-Abhath* 22: 3-22.

EI^{2}
Encyclopaedia of Islam. 2nd ed. Leiden and London: E.J. Brill.

Estori ha-Parchi
1897-9 *Caftor va-Pherach*, ed. A.M. Luncz. Jerusalem: Luncz Press (Hebrew).

Fraenkel, Y.
1992 The Endowing of the al-Madrasa al-Salahiyya Religious Foundation (Waqf) by Saladin. Pp. 64-85 in *Palestine in the Mamluk Period*, ed. J. Drory. Jerusalem: Yad Izhak Ben-Zvi (Hebrew).
1995 Rural Society in Mamluke Palestine. *Cathedra* 77:17-38. (Hebrew).

Franken, H.J., and Kalsbeek, J.
1975 Potters of a Medieval Village in the Jordan Valley. *Studies in Archaeology* 3:143-154.

von Groote, E.
1860 *Die Pilgerfahrt des Ritters Arnold von Harff.* Cologne: Steven's Druckerei.

Hamarneh, S.
1977-8 Sugarcane Cultivation and Refining under the Arab Muslims during the Middle Ages. *Annual of the Department of Antiquities of Jordan* 22:12-19.
1978 Medicinal Plants, Therapy and Ecology in al-Ghazzi's Book on Agriculture. *Studies in History of Medicine* 2:233-63.

Heyd, W.
1936 *Histoire du Commerce du Levant au Moyen-Age* II. Leipzig: Otto Harrassowitz.

Ibn Iyas, M.
1960-85 *Badai al-zuhur fi waqai al-duhur* I-IV. Cairo: Franz Steiner. (Arabic).

Ibn Tulun
1964 *Mufakhat al-Khilan fi Hawadith al-Zaman* I. Cairo: publisher not cited. (Arabic).

Ibrahim M.; Sauer J.A.; and Yassine, K.Y.
1976 The East Jordan Valley Survey, 1975 *Bulletin of the American Schools of Oriental Research* 222:41-66.

al-Idrisi, M.
1974 *Opus Geographicum* IV, eds, E. Cerulli, F. Gabrielli, G. Levi-Della, L. Petech, and G. Tucci. Napoli-Roma: Prostat apud E.J.Brill, Lugduni Batavorum.

Inalcik, H.
1970 The Heyday and Decline of the Ottoman Empire. Pp. 324-53 in *The Cambridge History of Islam* I. Ed. P.M. Holt and A.K.S. Lambton. Cambridge: Cambridge University Press.

Irwin, R.
1977 Iqta' and the End of the Crusader States. Pp. 62-77 in *The Eastern Mediterranean in the Period of the Crusades*, ed. P.M. Holt. Warminster: Aris and Phillips Press.

Laurent, J.C.M.
1864 Burchardi de Monte Sion Descriptio Terrae Sanctae. Pp. 1-95 in *Peregrinatores Medii Aevi Quatuor*. Leipzig: publisher not cited

Letts, M.
1953 *Mandeville's Travels, Texts and Translations* I. London:The /Hakluyt Society Press.

al-Maqrizi, A.
1911 *Ai-mawa'iz wal-i'tibar fi dhikr al-kitat wal-athar* I. Cairo: Bulak. (Arabic).

al-Maqrizi, A.
1973 *Kitab al-Suluk Ii-Ma'rifat Duwal al-Muluk,*II/2. Cairo: Bulak. (Arabic).

Mujir al-Din, A.
1973 *al-Uns al-Jalil bi-Tarikh al-Quds wa-al Khalil* II. Amman: publisher not cited. (Arabic).

al-Muqaddasi, S.M.
1906 *Ahsan al-taqasim fi marifat al-aqalim.* ed. de Goeje. Leyden: E.J. Brill (Arabic).

Newett, M.M.
1907 *Canon Pietro Casola's Pilgrimage to Jerusalem in the Year 1494.* Manchester: Manchester University Press.

Poliak, A.N.
1939 *Feudalism in Egypt, Syria, Palestine and the Lebanon, 1250-1900.* London: Royal Asiatic

Society.
1940 *Toldot ha'Yahasim ha'Karkai'yim b'Mizratim Suriya v'Eretz Yisrael b'Sof Y'mei ha'Bena'yim u'Bizman ha'Hadash* (The History of Soil Relations in Egypt, Syria, and the Land of Israel at the End of the Middle Ages and in the Modern Period). Jerusalem: 'Ahiever Press. (Hebrew).

Palestine Pilgrims' Text Society
1890 *The Library of the Palestine Pilgrims' Text Society.* London: Palestine Pilgrims' Text Society.

al-Qalqashandi, S.
1915 *Subh al-a'sha fi sina'at al insha',* IV-V, Cairo: publisher not cited (Arabic).

Rauwolf L.
1971 *Aigentliche Beschreibung der Raiss inn die Morgenlaender* (facsimile photocopy of 1583 edition). Graz: Akademische Druck.

Rohricht, R.
1895 Le Pelerinage du Moine Augustin Jacques de Verone (1335). *Revue de l'Orient Latin* 3:155-302.

Shoshan, B.
1992 Jerusalem Scholars (Ulama) and their Activities in the Mamluk Empire. Pp. 86-97 in *Palestine in the Mamluk period,* ed. J. Drory. Jerusalem: Yad Izhak Ben-Zvi (Hebrew).

Suriano Fra Francesco
1949 *Treatise on the Holy Land,* eds. T. Bellorini, E. Hoade and B.Bagatti. Jerusalem: Franciscan Printing Press.

Tov –'Alem, J.
1930 *Tzofnat Pa'aneah* II, ed. D. Herzog. Berlin: Josef Fischer. (Hebrew).

al-Umari, Ibn Fadlallah
1985 *Masalik al-absar fi mamalik al-amsar,* ed. A.F.Sayyid. Paris: Textes Arabes et Etudes Islamiques, 23. Cairo: Institute Française d'Archaéologie Orientale du Caire (Arabic).

Van-Berchem, M.
1909 *Materiaux pour un Corpus Inscriptionum Arabicarum* II/1*: Syrie du Nord.* Paris: Imprimerie de l'Institut Francais d'Archaéologie Orientale.

Wansbrough, J.
1965 Venice and Florence in the Mamluk Commercial Privileges. *Bulletin of the School of Oriental and African Studies* 28: 481-523.

Watson, A.M.
1983 *Agricultural Innovation in the Early Islamic World.* London and New York: Cambridge University Press.

Ya'ari, A. ed.
1943 *Letters of the Land of Israel.* Tel Aviv: Gazit (Hebrew).

Yaqut, Ibn Abd Allah al-Hamawi
1866-70 *Jacut's Geographisches Worterbuch,* ed. F.Wustenfeld. Leipzig: F.A.Brockhaus.

www.ingramcontent.com/pod-product-compliance
Lightning Source LLC
LaVergne TN
LVHW071213210926
848482LV00015B/806